DYNAMICS AND CONTROL OF HYBRID MECHANICAL SYSTEMS

WORLD SCIENTIFIC SERIES ON NONLINEAR SCIENCE

Editor: Leon O. Chua
University of California, Berkeley

WORLD SCIENTIFIC SERIES ON
NONLINEAR SCIENCE   Series B   Vol. 14

Series Editor: Leon O. Chua

DYNAMICS AND CONTROL OF HYBRID MECHANICAL SYSTEMS

edited by

Gennady Leonov
St Petersburg State University, Russia

Henk Nijmeijer
Eindhoven University of Technology, The Netherlands

Alexander Pogromsky
Eindhoven University of Technology, The Netherlands

Alexander Fradkov
Institute for Problems of Mechanical Engineering, St Petersburg, Russia

NEW JERSEY • LONDON • SINGAPORE • BEIJING • SHANGHAI • HONG KONG • TAIPEI • CHENNAI

Published by

World Scientific Publishing Co. Pte. Ltd.

5 Toh Tuck Link, Singapore 596224

USA office: 27 Warren Street, Suite 401-402, Hackensack, NJ 07601

UK office: 57 Shelton Street, Covent Garden, London WC2H 9HE

British Library Cataloguing-in-Publication Data
A catalogue record for this book is available from the British Library.

DYNAMICS AND CONTROL OF HYBRID MECHANICAL SYSTEMS
World Scientific Series on Nonlinear Science, Series B — Vol. 14

ISBN-13 978-981-4282-31-4
ISBN-10 981-4282-31-6

Printed in Singapore.

Dedicated to Ilya Izrailevich Blekhman
on the occasion of his 80th birthday

Preface

The book is based on the material presented at a mini-symposium "Dynamics and Control of Hybrid Mechanical Systems" at the 6th European Nonlinear Dynamics Conference (ENOC) held in St Petersburg, Russia, in 2008. In turn, the abovementioned mini-symposium was based on results of a similarly entitled Dutch-Russian research project funded by the Dutch organization for Pure Research (now) and the Russian Foundation for Basic Research. The project "Dynamics and Control of Hybrid Mechanical Systems" (DyCoHyMS) ran over the period 2006–2008 and turned out to be quite successful in terms of cooperation and scientific output. This is partly reflected in this book. A number of other related contributions were included into the volume and it now contains results of several international and interdisciplinary collaborations in the field, and reflects state-of-the-art scientific and technological development in the area of hybrid mechanical systems.

The papers in this volume aim to provide a better understanding of the dynamics and control of a large class of hybrid dynamical systems that are described by possibly different models in different state space domains. They not only cover important aspects and tools for hybrid systems analysis and control, but also a number of experimental realizations. Special attention is given to synchronization — a universal phenomenon in nonlinear science that gained tremendous significance since its discovery by Huijgens in the 17th century, see chapter 1 for an introduction to the observations of Huijgens regarding the phase synchronization of pendulum-clocks. Possible applications of the results introduced in the book include control of mobile robots, control of CD/DVD players, flexible manufacturing lines, and complex networks of interacting agents or robots.

It is our honor and pleasure to dedicate this book to Ilya Izrailevich

Blekhman, on the occasion of his 80th birthday celebrated in 2008. Professor Blekhman is one of the most profound thinkers and contributors in the area of nonlinear oscillations and synchronization in the XXth and XXIth century. His biography, for the first time published in English in such detail, follows below. We take the chance to wish Ilya Izrailevich Blekhman good health and new scientific achievements.

Gennady Leonov,
Henk Nijmeijer,
Alexander Pogromsky,
Alexander Fradkov

Biography: Ilya Izrailevich Blekhman

Ilya Izrailevich Blekhman, a leading Russian expert in the theory of nonlinear oscillations, the dynamics of machines and vibration technology was born on 29 November 1928 in Kharkov (now in Ukraine) and completed five years of secondary school in Leningrad (now Saint Petersburg) before World War II. During the blockade of the city in 1942 he was evacuated to the Urals, to Sverdlovsk (now Ekaterinburg) and, after passing two intermediate examinations as an external candidate, graduated from school with a gold medal award. Between 1945 and 1947 he studied at the Mechanics Department of the Urals Polytechnic Institute and, at the same time, at the Mechanics and Mathematics Faculty of the Urals State University as an external student. In 1947 he entered the Physics and Mathematics Faculty of Leningrad Polytechnic Institute, graduating with distinction as a research engineer in technical mechanics in 1951. Among the famous physicists, mathematicians and mechanics researchers whose lectures he attended during his student years were G.Yu. Dzhanelidze, A.F. Ioffe, A.I. Lurie, I.M. Malkin, B.V. Numerov, V.I. Smirnov and G.M. Fikhtengol'ts.

Blekhman began his research and engineering career in 1949 (still as a student) at the All-Union Scientific Research and Design Institute for Mechanical Processing of Mineral Resources (now Mekhanobr-Tekhnika Corporation), which he subsequently set up and is still working for the last 60 years. Currently he directs a department in the area of mechanics and applied mathematics. Since December 1996, he has been director of the laboratory of vibration mechanics run jointly by the Mekhanobr-Tekhnika Corporation and the Institute of Problems in Mechanical Engineering of the Russian Academy of Sciences. He defended his candidate (Ph.D.) dissertation in 1955 under the supervision of Prof. A.I. Lurie and his doctoral thesis in 1962. He got the title of professor in 1969.

Blekhman has promoted and developed several new areas in the theory of nonlinear oscillations, in applied mechanics, and in the foundations of vibration processes and mechanical engineering. Among his most important achievements are:

- the development (together with G.Yu. Dzhanelidze) of the theory of vibrational motion — the process of "directed" slow changes, which occur under the effect of rapid "undirected actions";
- the discovery of self-synchronization of rotating bodies (rotors) and the creation of a theory to explain this phenomenon;
- the determination of an extremum condition for stability, which extends the classical Lagrange-Dirichlet theory of the stability of equilibrium positions to synchronized rotations of weakly interacting bodies;
- the discovery and investigation of a class of nonlinear mechanical systems which, during vibration, acquire the "average potential" property: slower motions in those systems correspond to the motions of a certain average potential system, although the original system is highly non-conservative;
- the application of the classical Laval self-balancing principle to multi-rotor and nonlinear systems;
- the discovery (jointly with colleagues) and theoretical justification of the development of gravitational gas lift flows, which is useful for the efficient exploitation of gas lift flows, which has facilitated plans for the efficient exploitation of the energy and raw materials from the Pacific Ocean and which also contributed to the prevention of some ecological disasters;
- the development and validation of a general mechanical-mathematical approach to study of the effect of vibration on nonlinear mechanical systems (vibration mechanics and vibrorheology);
- introduction (jointly with K.A. Lurie) of a new principle in material science: the idea of dynamic materials and composites.

Many of these results have been presented in reference publications and textbooks. He is an author or co-author of more than 150 scientific papers, 2 scientific registered discoveries, 40 inventions and patents and 8 books, including "Vibrational Transportation" (with G.Yu. Dzhanelidze, Moscow: Nauka, 1964, in Russian), "Synchronization of dynamical systems" (Moscow: Nauka, 1971, in Russian), "Synchronization in science and technology" (New-York: ASME Press, 1988), "Vibrational mechanics" (Singapore: World Scientific, 2000; in Russian: Moscow, Fizmatlit, 1994). He is a co-author, with A. D. Myshkis and Ya. G. Panovko, of a methodological book "Applied Mathematics: Subject, Logic and Features" (published in Russia in 1976, 1991 and 2005), which describes methodological features of applied mathematics as a science and which has received wide recognition. The discovery, theoretical justification and description of the synchronization of rotating bodies and the development of gravitational gas lift flows have been registered as scientific discoveries.

Blekhman has made some important contributions in engineering designs, notably, the construction of new vibration machines for enriching mineral resources-crushers, mills, sifters, flotation units, concentrators and so on. He and his successors have taken out many patents, and have sold licenses to leading firms in the USA, England and Germany.

Blekhman has established a leading Russian research school in the area of the theory of vibration processes and machines; seven Doctor of Sciences and about 40 Candidate of Sciences (Ph.D.) degrees have been awarded under his supervision. He is a member of panels which judge doctoral dissertations at the Mekhanobr-Tekhnika Corporation, Institute of Problems in Mechanical Engineering and the St. Petersburg Marine Technical University, as well as a member of a number of research councils of the Russian Academy of Sciences. He has given several courses of lectures for post-graduates and undergraduates in various higher technical teaching institutes in St. Petersburg and has lectured at MIT and Worcester Polytechnic in the USA, at polytechnic universities in Great Britain, Germany, the Netherlands, Denmark and Poland and at a number of other universities and institutes in other countries. Blekhman has been a member of Russian National Committees on theoretical and applied mechanics and (since 1965) on the theory of machines and mechanisms; in 1990 he was elected a full member of the Russian Engineering Academy. Blekhman was a member of editorial boards of the journals *Applied Mathematics and Mechanics* and *Mechanics of Solids* for over more than 50 years.

In 1998 Blekhman was awarded the Prize of the Russian Government in Science and Technology, in 1999 he received the Alexander von Humbolt Prize (Germany), and in 2000 the Al-Khoresmi Prize (Iran). In 2001 he received the honorary title "Honorary Machine Builder of Russian Federation", in 2003 he received the honorary title "Distinguished Reasercher of the Russian Federation". In 2009, Blekhman was awarded the Tchebyshev Prize in the area of mathematics and mechanics.

Contents

    Variational Inequalities   29

    G.A. Leonov, V. Reitman

    3.1 Introduction 29
    3.2 Evolutionary variational inequalities 31
    3.3 Basic assumptions 33
    3.4 Absolute observation - stability of evolutionary
         inequalities 34
    3.5 Application of observation stability to the beam
         equation 37

4. A Discrete-time Hybrid Lurie Type System   43

    V.N. Belykh, B. Ukrainsky

    4.1 Introduction 43
    4.2 Reduction to a normal form 44
    4.3 Existence of invariant domain 45
    4.4 Conditions of hyperbolicity 48

5. Frequency Domain Performance Analysis of Marginally
    Stable LTI Systems with Saturation   53

    R.A. van den Berg, A.Y. Pogromsky, J.E. Rooda

    5.1 Introduction 53
    5.2 LTI system with saturation 55
         5.2.1 System description 55
         5.2.2 Case: electromechanical system 57
         5.2.3 Motivating example: nonlinear behavior 58
    5.3 Convergent systems and simulation-based frequency
         domain analysis 60
         5.3.1 Convergent systems 60
         5.3.2 Convergent system design 62
         5.3.3 Performance analysis in frequency domain . . . 62
    5.4 Frequency domain analysis based on describing function
         approach 65
         5.4.1 Describing function method 65
         5.4.2 Performance analysis example 68
    5.5 Conclusion 69

9. Stability and Control of Lur'e-type Measure Differential Inclusions   129

N. van de Wouw, R. I. Leine

10. Synchronization between Coupled Oscillators: An Experimental Approach   153

D.J. Rijlaarsdam, A.Y. Pogromsky, H. Nijmeijer

     and Synchronization   211

     A.L. Fradkov, B.R. Andrievskiy, K.B. Boykov, B.P. Lavrov

     14.1   Introduction 211
     14.2   Design of mechanical part 213
     14.3   Electronics of the multipendulum setup 214
            14.3.1   System for data exchange with control
                     computer 215
            14.3.2   Architecture of the data exchange system 216
            14.3.3   Computer-process interface 216
            14.3.4   Electronic modules of the set-up 217
            14.3.5   Communications protocol 218
     14.4   Conclusions 218

15. High-frequency Effects in 1D Spring-mass Systems
     with Strongly Non-linear Inclusions   223

     B.S. Lazarov, S.O. Snaeland, J.J. Thomsen

     15.1   Introduction 223
     15.2   Mechanical model 227
     15.3   Band gap effects in periodic structures 228
     15.4   Approximate equations governing the slow and the fast
            motion 229
            15.4.1   Example 1: linear plus cubic non-linearity . . . 233
            15.4.2   Example 2: essentially non-linear damping and
                     restoring forces 233
     15.5   Numerical examples 234
            15.5.1   Inclusions with linear plus cubic non-linear
                     behaviour 235
            15.5.2   Non-linear inclusions with non-local
                     interaction 236
            15.5.3   Linear chains with non-linear damping forces . . 237
     15.6   Conclusions 239

Huijgens' Synchronization: A Challenge

H. Nijmeijer, A.Y. Pogromsky

Department of Mechanical Engineering
Eindhoven University of Technology
P.O. Box 513, 5600 MB Eindhoven, The Netherlands

Oscillations are common almost everywhere, be it in biology, in economics, in physics and many other fields. Everyone is familiar with the day-night rhythm, or the regular or less regular heart-beat of a human, the pig cycle in economy or the flashing of fire-flies and so on. All the above examples have in common that the oscillations seem to happen naturally, but there are also other more or less forced type of oscillations like for instance in chemistry, electrical circuits and acoustics. Probably the most basic example of an oscillator is a pendulum clock that runs at a fixed frequency and such that the exact time is given by the clock. Design and construction of a fully accurate mechanical clock is – even today – a very challenging task; the reader is referred to [Rawlings, 1994; Penman, 1998] for some background on this.

Sofar we haven't defined what exactly an oscillation is, and for the time being it suffices to understand it as a dynamic feature that "repeats" itself in time, usually with a fixed time period. As mentioned, such oscillations are abundant in every day life, but even more remarkable, in biological systems the "biological clock" is – or becomes – identical. The flashing of a flock of fire-flies may serve as a good example of this. In fact, there is a large literature on this "synchronization", i.e. "conformity in time" in biological systems (and in some cases this conformity in time agrees with the earth time clock of 24 hours a day). A nice account on synchronization in – amongst others – biology can be found in [Strogatz, 2003]. Synchronization, sometimes also referred to as coordination or cooperation, also happens in other than biological systems. Again many examples exist, but an illuminating example is dancing at a music festival – as should be clear the music made by orchestra already forms an illustration of synchronization – where the dancers' motion synchronizes with the rhythm of the music.

There are numerous examples of synchronization reported in the literature, but here we will in particular only focus on some of the earlier reported cases. For a more complete account on synchronization *per se* we refer to [Pikovsky *et. al.*, 2001]. Probably the earliest writing on synchronization is due to Huijgens, see [Huygens, 1660] and describes the (anti-phase) synchronization of a pair of pendulum clocks. Christiaan Huijgens (1629-1695) is a famous Dutch scientist who worked on subjects from astronomy, physics and mathematics. He was famous for his work in optics and the construction of (pendulum) clocks and telescopes. Huijgens describes his observation that the two pendulum clocks as described in Figure 1.1 exhibit (mirrored) synchronized motion, even when they are initiated at different initial conditions, see [Huygens, 1660], or for an English translation the reader may consult [Pikovsky *et. al.*, 2001]. This is by Huijgens called "sympathy of two clocks", who also ingeniously linked this to the beam to which the two clocks are attached.

Fig. 1.1   Clocks attached to a beam on two chairs.

Indeed, the unprecedented contribution at this point is the correct explanation by Christiaan Huijgens for the (anti-) phase synchronization of the two pendulum clocks. The reader should realize that Huijgens lived in an era that differential calculus did not exist yet – or rather Huijgens himself was one of the earliest initiators of it, whereas Newton and Leibnitz only at the end of the 17th century, beginning 18th century, fully developed the concepts from differential calculus. Therefore, as in the preceding centuries, see e.g. the work of another Dutch (or rather Flemish) scientist, Simon Stevin, the best one can do expect in studying new phenomena, is the combination of "Spiegheling ende Daet", i.e. reflection and experiment, [Devreese and van den Berghe, 2003]. This is exactly what Huijgens did regarding his two pendulum clock experiment as described in Figure 1.1, namely he repeated the study about phase synchronization of the two pendula under systematically changing the set-up, and most notably, by changing the distance between the two clocks hanging on the beam. He then also reached the conclusion that once the clocks are positioned close to each other, in phase synchronization will occur. Huijgens realized the importance of his discovery as this would allow ships sailing at the ocean to obtain a second "position" measurement, besides the relative position of the sun, so that the ship can determine its position at the ocean.

Today, several experiments mimicking the Huijgens' experiment have been reported, see e.g. [Panteleone, 2002; Benett *et. al.*, 2002], all aiming at demonstrating that in-phase or anti-phase synchronization of a number of pendulum-like oscillators can be achieved. In-phase synchronization in a Huygens-type setup was explained in [Blekhman, 1998] by means of a model of coupled Van der Pol equations.

At this point it is worth noting that many of the reported toy-experiments of Huijgens' synchronization are a simplification of the work done by Huijgens! Mathematically, the experiment as reported by Huijgens can be cast into a set of differential equations. Namely, for each of the pendulum clocks a simple oscillator equation suffices – at least under the assumption that the so-called escapement mechanism that keeps the clocks running, can be modeled in a simple manner – and in addition the beam to which the clocks are attached needs to be modeled by a partial differential equation with suitable boundary conditions. Thus, the set-up consisting of two pendula linked to a free hanging beam is automatically an infinite dimensional system, for which a rigorous study of the in-phase or anti-phase synchronization of the two pendula is, as far as is known, still never addressed in the literature. Such a study, albeit dealing with the (partial) stability of an infinite dimensional system, typically is expected to involve a study of the energy of the overall system, and thus some kind of Lyapunov (or infinite dimensional extensions like the Lyapunov-Krasovskii functional

approach) seems in order. So far, most of the studies dealing with the Huijgens' synchronization problem treat a finite dimensional approximation,
e.g. in an earlier attempt [Oud et.al., 2006] we proposed a 3 degrees of
freedom approximation of the Huijgens' set-up, see Figure 1.2, which to
some extent does capture the effect that energy from one pendulum can be
transferred to the other via the swinging platform.

Fig. 1.2   A setup with two metronomes.

In the recent paper [Dilao, 2009] a slightly different approach is given
to arrive at a finite dimensional (4 dof) model mimicking partly the Huijgens'experiment. Again, in this case a fairly complete analysis on in phase
and anti-phase synchronization of the oscillators is possible.

Numerical results related to the 'true' infinite dimensional Huijgens'
problem illustrating the in phase and anti-phase synchronization may form
a next step, compare this to [Czolczynski et. al., 2007]. Thus, besides several experimental illustrations and simulation studies, a rigorous analysis
regarding the synchronization of pendula is still pending. For a thorough
understanding of Huijgens' synchronization one can essentially distinguish
two challenges. On the one hand, this is the issue of determining all the
possible "stationary" solutions (like the in-phase and anti-phase pendulum clocks) and secondly, a stability-study regarding these "stationary"
solutions. However, the old Huijgens' synchronization problem, and in particular its analysis is even today a very timely research theme. It centers
about advanced nonlinear dynamics, hybrid dynamics – given the impulsive nature of the escapement mechanism – and in particular regarding the
synchronization of multiple clocks, energy-based control techniques come
into play. The contributions in this book largely fit into this broad arena of
exciting dynamical systems. On one hand, there are several contributions

that aim at a complete stability analysis for the error dynamics between various kinds of oscillators, coupled through a relatively simple proportional error term. In many cases, energy and Lyapunov-like arguments are developed towards a successful stability theory.

On the other hand, several authors have contributed towards the study of all potential limiting or stationary solutions. Despite the developed results in this book, a complete and rigorous proof of the in-phase respectively anti-phase synchronization such as observed and described by Huijgens more than three centuries ago is still missing. However, it is strongly felt that the contributions given here may help in a further understanding of synchronization of pendulum clocks and more generally other oscillators. It is clear that for a successful analysis for such situations besides the particular type of oscillators also the selected type of coupling is of key importance. It is our belief that the results given here provide important and illuminating insight in the old and intriguing synchronization problem. In addition, it may look artificial in the beginning, but the actual "sympathy" between pendulum clocks inherently contains both in experiments as well in analysis typical hybrid aspects that tremendously complicate the problem. In that sense, Dynamics and Control of Hybrid Mechanical Systems, remains a very exciting and challenging field of research.

References

Bennett, M., Schat, M., Rockwood, H. and Wiesenfeld, K. (2002) Huygens's clocks, *Proceedings of the Royal Society A: Mathematical, Physical and Engineering Sciences* 458(2019), pp. 563–579.

Blekhman, I.I. (1998) *Synchronization in science and technology*, ASME. New York.

Czolczynski, D.K., Stefanski, A., Perlowski, P. and Kapitaniak, T. (2007) "Periodization and synchronization of duffing oscillators suspended on elastic beam", in H.Y.Hu and E.Kreuzer (eds). *IUTAM Symposium on Dynamics and Control of Nonlinear Systems with Uncertainty*, pp.317-322.

Devreese J.T. and van den Berghe, G. (2003) "Wonder en is gheen wonder", de geniale wereld van Simon Stevin 1548-1620, Davidsfonds/Leuven (in Dutch).

Dilao, R. (2009) "Antiphase and in-phase synchronization of nonlinear oscillators: the Huygens's clock system", *Chaos* **19**, 23118.

Huygens, C. (1660), *Oeuvres complétes de Christiaan Huygens*, Martinus Nijhoff, vol. 5,17, 1893, 1932.

Oud W., Nijmeijer, H. and Pogromsky A. (2006), A study of Huygens synchronization. Experimental Results, *Proceedings of the 1st IFAC Conference on*

Analysis and Control of Chaotic Systems, France, CDROM.

Pantaleone J. (2002), Synchronization of metronomes, *American Jounal of Physics*, 70(10), pp. 992–1000.

Penman, L. (1998), *Practical Clock Escapements*, Mayfield Books UK and Clock-Works Press US.

Pikovsky, A., Rosenblum, M. and Kurths, J. (2001) *Synchronization, a universal concept in nonlinear sciences*, Cambridge Nonlinear Science Series 12.

Rawlings, A.L. (1994) *The Science of Clocks and Watches*, Third edition, The British Horological Institute, Upton Hall, Upton Newark.

Strogatz, S. (2003) *SYNC. The engineering Science of Spontaneous Order*, Hyperion, New York.

Lyapunov Quantities and Limit Cycles of Two-dimensional Dynamical Systems

N.V. Kuznetsov, G.A. Leonov

Department of Mathematics and Mechanics
St. Petersburg State University
Russia

Abstract

This paper is devoted to the computation of Lyapunov quantities (Poincare-Lyapunov constants) and to the study of limit cycles of two-dimensional dynamical systems.

Here new method for computation of Lyapunov quantities is suggested, which is based on the constructing approximations of solution in the original Euclidean coordinates and in the time domain. The advantages of this method are in its ideological simplicity and visualization power. This approach can also be applied to the problem of distinguishing of isochronous center.

New method of asymptotic integration for Lienard equation is suggested which is effective for localization of attractors and investigation of large limit cycles. This technique is applied for investigation of autonomous quadratic systems. The quadratic system is reduced to the Lienard equation and then the two-dimensional domain of parameters, corresponding to the existence of four limit cycles (three small and one large), is evaluated. This domain extends the domain of parameters, obtained for quadratic system with four limit cycles by S.L. Shi in 1980.

2.1   Introduction

In the classical works of Poincare [Poincare, 1885] and Lyapunov [Lyapunov, 1892], for the study of system behavior in the critical case (the case of a pair of pure imaginary roots of linearized system), the method of analysis of the so-called Lyapunov quantities (or Poincare-Lyapunov constants) was developed. The computation of Lyapunov quantities is closely connected with the important in engineering mechanics question of dynamical system behavior near the boundary of stability domain. Followed by the work of Bautin [Bautin, 1949], one differs safe or unsafe boundaries, a slight shift of which implies a small (invertible) or noninvertible changes of system behavior, respectively. Such changes correspond, for example, to scenario of "soft" or "hard" excitations of oscillations, considered by Andronov [Andronov *et al.*, 1966].

This method, following the famous works of Bautin [Bautin, 1939, 1952], permits one to study effectively bifurcation of small limit cycles (or large limit cycles in the neighborhood of infinity).

The study of limit cycles of two-dimensional dynamical systems was stimulated by purely mathematical problems (Hilbert's sixteenth problem, the center-and-focus problem) as well as by many applied problems such as the oscillations of electronic generators and electrical machines, the dynamics of populations, and the safe and unsafe boundaries of stability (see, e.g., [Bautin and Leontovich, 1976; Andronov *et al.*, 1966; Anosov *et al.*, 1997; Roussarie, 1998; Shilnikov *et al.*, 2001] and many others).

In 1901 year Hilbert, in its famous 16th problem, has set a problem of study of hidden oscillations: the analysis of mutual disposition and the number of limit cycles for two-dimensional polynomial systems. In a time more than century, within the framework of solution of this problem the numerous theoretical and numerical results are obtained (see bibliography in [Reyn, 1994]) but the problem is still far from being resolved even for simple classes of systems. For example, the problem of determining quadratic systems, in which there is a limit cycle, is nontrivial and rather difficult. V.I. Arnold writes (translated from [Arnold, 2005]): "To estimate the number of limit cycles of square vector fields on plane, A.N. Kolmogorov had distributed several hundreds of such fields (with randomly chosen coefficients of quadratic expressions) among a few hundreds of students of Mechanics and Mathematics Faculty of MGU as a mathematical practice. Each student had to find the number of limit cycles of his/her field. The result of this experiment was absolutely unexpected: not a single field had a limit cycle! It is known that a limit cycle persists under a small change of field coefficients. Therefore, the systems with one, two, three (and even, as has become known later, four) limit cycles form an open set in the space of coefficients and, therefore, for a random choice of polynomial coefficients, the probability of hitting in it is positive. The fact that this did not occur suggests that the above-mentioned probabilities are, apparently, small."

In the present work the new effective methods of computation of Lyapunov quantities and localization and study of limit cycles and the domain of existence of limit cycle in the space of parameters will be demonstrated.

2.2 Computation of Lyapunov quantities and small limit cycles

In studying small limit cycles in the neighborhood of equilibrium of two-dimensional dynamical systems, one of the central problems is the computation of Lyapunov quantities (see, e.g, [Bautin, 1949; Cherkas, 1976; Marsden and McCracken, 1976; Lloyd, 1988; Li, 2003; Dumortier *et. al.*, 2006; Christopher and Li, 2007] and others).

Consider a system of two autonomous differential equations

$$\frac{dx}{dt} = -y + f(x, y),$$
$$\frac{dy}{dt} = x + g(x, y),$$
(2.1)

where $x, y \in \mathbb{R}$. Here the functions $f(\cdot, \cdot)$ and $g(\cdot, \cdot)$ are sufficiently smooth and their expansion begins with the terms of at least the second order. Therefore, there are two purely imaginary eigenvalues of linear part of system (2.1).

At present, there exist different methods for determining Lyapunov quantities and the computer realizations of these methods, which make it possible to obtain Lyapunov quantities in the form of symbolic expressions, depending on expansion coefficients of the right-hand sides of equations of system (see, e.g., [Yu and Chen, 2008]). These methods differ in complexity of algorithms, compactness of obtained symbolic expressions, and also in a space in which computations are performed.

While the general expression for the first and second Lyapunov quantities (in terms of expansion coefficients of f and g in the original space) had been computed in the 40-50s of last century [Bautin, 1949; Serebryakova, 1959], the third Lyapunov quantity was computed only in 2008 [Kuznetsov and Leonov, 2008a; Leonov *et. al*, 2008]. Its expression occupies more than four pages and the expression for the fourth Lyapunov quantity occupies more than 45 pages. Therefore the analysis of general expressions of Lyapunov quantities is a difficult algebraic problem.

The first method for the computation of Lyapunov quantities was suggested by Poincare [Poincare, 1885] and Lyapunov [Lyapunov, 1892]. This method consists in sequential obtaining time-independent integrals for approximations of system. Since $V_2(x,y) = \dfrac{(x^2 + y^2)}{2}$ is an integral of the first approximation system and the system (2.1) is sufficiently smooth, then in a certain small neighborhood of zero state we can sequentially construct the Lyapunov function of the form

$$V(x,y) = \frac{x^2 + y^2}{2} + V_3(x,y) + \dots + V_k(x,y), \qquad (2.2)$$

where $V_k(x,y) = \sum\limits_{i+j=k} V_{i,j} x^i y^j$ are homogeneous polynomials with the unknown coefficients $\{V_{i,j}\}_{i+j=k,\ i,j\geqslant 0}^k$. These coefficients can be chosen sequentially (via the coefficients of expansions of the functions f and g and the coefficients $\{V_{i,j}\}_{i+j<k}$, determined at the previous steps of iteration) in such a way that the derivative of $V(x,y)$ by virtue of system (2.1)

$$\dot{V}(x,y) = \frac{\partial V(x,y)}{\partial x}(-y + f(x,y)) + \frac{\partial V(x,y)}{\partial y}(x + g(x,y)) \qquad (2.3)$$

takes the form

$$\dot{V}(x,y) = w_1(x^2 + y^2)^2 + w_2(x^2 + y^2)^3 + \dots + o\big((|x| + |y|)^{k+1}\big), \qquad (2.4)$$

where coefficients w_i depend only on the coefficients of expansions of f and g.

Sequentially determining the coefficients of the form V_k for $k = 3, \dots$ (for which purpose at each step it is necessary to solve a system of $(k+1)$ linear equations), from (2.3) and (2.4) it may be obtained the coefficient w_m that is the first not equal to zero coefficient:

$$\dot{V}(x,y) = w_m(x^2 + y^2)^{m+1} + o\big((|x| + |y|)^{2m+2}\big).$$

The coefficient w_m is usually called [Chavarriga and Grau, 2003] a Poincare-Lyapunov constant ($2\pi w_m$ is mth Lyapunov quantity [Frommer, 1934]). By the additional conditions [Lynch, 2005]

$$V_{2m,2m+2} + V_{2m+2,2m} = 0, \quad V_{2m,2m} = 0$$

at the kth step of iteration the coefficients $\{V_{i,j}\}_{i+j=k}$ can be determined uniquely.

Another approach to the computation of Lyapunov quantities involves the determination of approximations of solution of system. For this purpose, in a classical approach [Lyapunov, 1892] it is used the changes of variables to reduce "turn" time of all trajectories to a constant and procedures for recurrent construction of solution approximations.

Having performed the changes $x = r \cos \phi$, $y = r \sin \phi$, we obtain system (2.1) in the polar coordinates r and ϕ. It can be represented in the form

$$\frac{dr}{d\phi} = \frac{f(r \cos \phi, r \sin \phi) \cos \phi + g(r \cos \phi, r \sin \phi) \sin \phi}{1 - \dfrac{f(r \cos \phi, r \sin \phi) \sin \phi}{r} + \dfrac{g(r \cos \phi, r \sin \phi) \cos \phi}{r}} = R(r, \phi).$$

Here $R(0,0) = 0$ and for sufficiently small r the function $R(r, \phi)$ is smooth (due to the smoothness of f and g). Consider the representation

$$\frac{dr}{d\phi} = rR_1(\phi) + r^2 R_2(\phi) + \dots.$$

Then, substitute into this equation the following representation of solution with the initial date $(0, r_0)$:

$$r(\phi, 0, r_0) = u_1(\phi)r_0 + u_2(\phi)r_0^2 + \dots \quad u_1(0) = 1, \ u_{i>1}(0) = 0. \qquad (2.5)$$

We obtain the equations for sequential calculation of $u_i(\phi)$

$$\begin{aligned} \dot{u}_1(\phi) &= R_1(\phi), \\ \dot{u}_2(\phi) &= R_1(\phi)u_2(\phi) + R_2(\phi)u_1^2(\phi), \\ &\cdots \end{aligned} \qquad (2.6)$$

By the solution for $\phi = 2\pi$ we have

$$r = r(2\pi, 0, r_0) = \alpha_1 r_0 + \alpha_2 r_0^2 + \alpha_3 r_0^3 + \dots, \qquad \alpha_i = u_i(2\pi),$$

where α_i are called focus values. If $\alpha_2 = \dots = \alpha_{2m} = 0$, then the α_{2m+1} is called mth Lyapunov quantity.

Further, later on there were developed the different computing methods, involving the computation in complex space, the reduction of system to a normal form (see, e.g., [Yu, 1998; Li, 2003; Yu and Chen, 2008]), and special analytical-numerical methods [Gasull *et al.*, 1999].

2.2.1 *Computation of Lyapunov quantities in Euclidean coordinates and in the time domain*

In the present work a new method for computation of Lyapunov quantities is considered which is based on constructing approximations of solution (as a finite sum in powers of degrees of initial data) in the original Euclidean coordinates and in the time domain [Leonov *et al.*, 2008; Kuznetsov and Leonov, 2008b]. The advantages of this method are in its ideological simplicity and visualization power. This approach can also be applied to the problem of distinguishing of isochronous center [Chavarriga and Grau, 2003; Gine, 2007; Gasull *et al.*, 1997; Sabatini and Chavarriga, 1999] since it permits one to find an approximation of "turn" time of trajectory as a function of initial data.

2.2.1.1 *Approximation of solution in Euclidean coordinates*

Suppose, the functions $f(\cdot,\cdot)$ and $g(\cdot,\cdot)$ have continuous partial derivatives with respect to x and y up to $(n+1)$th order in the open neighborhood U of radius R_U at the point $(x,y) = (0,0)$:

$$f(\cdot,\cdot), g(\cdot,\cdot) \colon \mathbb{R} \times \mathbb{R} \to \mathbb{R} \quad \in \mathcal{C}^{(n+1)}(U). \tag{2.7}$$

We also assume that the first terms of their expansion are the terms of at least the second order, namely

$$f(x,y) = \sum_{k+j=2}^{n} f_{kj}x^k y^j + o\big((|x|+|y|)^n\big) = f_n(x,y) + o\big((|x|+|y|)^n\big),$$
$$g(x,y) = \sum_{k+j=2}^{n} g_{kj}x^k y^j + o\big((|x|+|y|)^n\big) = g_n(x,y) + o\big((|x|+|y|)^n\big). \tag{2.8}$$

The existence condition for $(n+1)$th partial derivatives of f and g with respect to x and y is only used for simplicity of exposition and can be weakened.

Further we will use a smoothness of the functions f and g and follow the first Lyapunov method on finite time interval [Lefschetz, 1957; Cesari, 1959].

Let $x(t,x(0),y(0))$, $y(t,x(0),y(0))$ be a solution of system (2.1) with the initial data

$$x(0) = 0, \ y(0) = h. \tag{2.9}$$

Denote

$$x(t,h) = x(t,0,h), \ y(t,h) = y(t,0,h).$$

Below a time derivative will be denoted by x' and $\dot{x}$.

Lemma 2.1. *The positive numbers $H \in (0,R_U)$ and $\Delta T > 0$ exist such that for all $h \in [0,H]$ the solution $\big(x(t,h),y(t,h)\big)$ is defined for $t \in [0,2\pi + \Delta T]$.*

The validity of lemma follows from condition (2.7) and the existence of two purely imaginary eigenvalues of the matrix of linear approximation of system (2.1).

This implies [Hartman, 1984] the following

Lemma 2.2. *If smoothness condition (2.7) is satisfied, then*

$$x(\cdot,\cdot), y(\cdot,\cdot) \in \mathcal{C}^{(n+1)}\big([0, 2\pi + \Delta T] \times [0, H]\big) \tag{2.10}$$

Now we consider sufficiently small initial data $h \in [0, H]$ and a finite time interval $t \in [0, 2\pi + \Delta T]$. Then we use a uniform boundedness of the solution $(x(t, h), y(t, h))$ and its mixed partial derivatives with respect to h and t up to the order $(n + 1)$ inc in the set $[0, 2\pi + \Delta T] \times [0, H]$.

Further we apply a well-known linearization procedure [Leonov and Kuznetsov, 2007]. From Lemma 2.2 it follows that for each fixed t the solution of system can be represented

$$x(t, h) = h \frac{\partial x(t, \eta)}{\partial \eta}\Big|_{\eta=0} + \frac{h^2}{2} \frac{\partial^2 x(t, \eta)}{\partial \eta^2}\Big|_{\eta=h\theta_x(t,h)} \quad 0 \le \theta_x(t, h) \le 1,$$

$$y(t, h) = h \frac{\partial y(t, \eta)}{\partial \eta}\Big|_{\eta=0} + \frac{h^2}{2} \frac{\partial^2 y(t, \eta)}{\partial \eta^2}\Big|_{\eta=h\theta_y(t,h)} \quad 0 \le \theta_y(t, h) \le 1.$$
$$\tag{2.11}$$

Note that by Lemma 2.2 and relation (2.11), the functions

$$\frac{h^2}{2} \frac{\partial^2 x(t, \eta)}{\partial \eta^2}\Big|_{\eta=h\theta_x(t,h)}, \quad \frac{h^2}{2} \frac{\partial^2 y(t, \eta)}{\partial \eta^2}\Big|_{\eta=h\theta_y(t,h)}$$

together with their time derivatives are smooth functions of t and have the order of smallness $o(h)$, uniform with respect to t, on a considered finite time interval $[0, 2\pi + \Delta T]$.

Introduce the following denotations

$$\widetilde{x}_{h^k}(t) = \frac{\partial^k x(t, \eta)}{k! \, \partial^k \eta}\Big|_{\eta=0}, \quad \widetilde{y}_{h^k}(t) = \frac{\partial^k y(t, \eta)}{k! \, \partial^k \eta}\Big|_{\eta=0}.$$

We shall say that the sums

$$x_{h^m}(t, h) = \sum_{k=1}^{m} \widetilde{x}_{h^k}(t) h^k, \quad y_{h^m}(t, h) = \sum_{k=1}^{m} \widetilde{y}_{h^k}(t) h^k$$

are the mth approximation of solution of system with respect to h. Substitute representation (2.11) in system (2.1). Then, equating the coefficients of h^1 and taking into account (2.8), we obtain

$$\frac{d\widetilde{x}_{h^1}(t)}{dt} = -\widetilde{y}_{h^1}(t), \quad \frac{d\widetilde{y}_{h^1}(t)}{dt} = \widetilde{x}_{h^1}(t). \tag{2.12}$$

Hence, by conditions on initial data (2.9) for the first approximation of the solution $(x(t, h), y(t, h))$ with respect to h, we have

$$x_{h^1}(t, h) = \widetilde{x}_{h^1}(t)h = -h\sin(t), \quad y_{h^1}(t, h) = \widetilde{y}_{h^1}(t)h = h\cos(t). \quad (2.13)$$

Similarly, to obtain the second approximation $(x_{h^2}(t, h), y_{h^2}(t, h))$, we substitute the expressions

$$x(t, h) = x_{h^2}(t, h) + \frac{h^3}{3!} \frac{\partial^3 x(t, \eta)}{\partial \eta^3} \Big|_{\eta = h\theta_x(t,h)},$$

$$(2.14)$$

$$y(t, h) = y_{h^2}(t, h) + \frac{h^3}{3!} \frac{\partial^3 y(t, \eta)}{\partial \eta^3} \Big|_{\eta = h\theta_y(t,h)}.$$

in formula (2.8) for $f(x, y)$ and $g(x, y)$. Note that by (2.8) in expressions for f and g, the coefficients of h^2 (denoted by $u_{h^2}^f$ and $u_{h^2}^g$, respectively) depend only on $\widetilde{x}_{h^1}(t)$ and $\widetilde{y}_{h^1}(t)$, i.e., by (2.13) the coefficients are known functions of time and are independent of the unknown functions $\widetilde{x}_{h^2}(t)$ and $\widetilde{y}_{h^2}(t)$. Then, we have

$$f\big(x_{h^2}(t, h) + o(h^2), y_{h^2}(t, h) + o(h^2)\big) = u_{h^2}^f(t)h^2 + o(h^2),$$
$$g\big(x_{h^2}(t, h) + o(h^2), y_{h^2}(t, h) + o(h^2)\big) = u_{h^2}^g(t)h^2 + o(h^2).$$

Substituting (2.14) in system (2.1), for the determination of $\widetilde{x}_{h^2}(t)$ and $\widetilde{y}_{h^2}(t)$ we obtain

$$\frac{d\widetilde{x}_{h^2}(t)}{dt} = -\widetilde{y}_{h^2}(t) + u_{h^2}^f(t),$$

$$(2.15)$$

$$\frac{d\widetilde{y}_{h^2}(t)}{dt} = \widetilde{x}_{h^2}(t) + u_{h^2}^g(t).$$

Lemma 2.3. *For solutions of the system*

$$\frac{d\widetilde{x}_{h^k}(t)}{dt} = -\widetilde{y}_{h^k}(t) + u_{h^k}^f(t),$$

$$(2.16)$$

$$\frac{d\widetilde{y}_{h^k}(t)}{dt} = \widetilde{x}_{h^k}(t) + u_{h^k}^g(t)$$

with the initial data

$$\widetilde{x}_{h^k}(0) = 0, \quad \widetilde{y}_{h^k}(0) = 0 \quad (2.17)$$

we have

$$\widetilde{x}_{h^k}(t) = u_{h^k}^g(0)\cos(t) + \cos(t)\int_0^t \cos(\tau)\big((u_{h^k}^g(\tau))' + u_{h^k}^f(\tau)\big)\,d\tau +$$

$$+ \sin(t)\int_0^t \sin(\tau)\big((u_{h^k}^g(\tau))' + u_{h^k}^f(\tau)\big)\,d\tau - u_{h^k}^g(t),$$

$$\widetilde{y}_{h^k}(t) = u_{h^k}^g(0)\sin(t) + \sin(t)\int_0^t \cos(\tau)\big((u_{h^k}^g(\tau))' + u_{h^k}^f(\tau)\big)\,d\tau -$$

$$- \cos(t)\int_0^t \sin(\tau)\big((u_{h^k}^g(\tau))' + u_{h^k}^f(\tau)\big)\,d\tau.$$

$$(2.18)$$

The relations (2.18) can be verified by direct differentiation.

Repeating this procedure for the determination of coefficients $\widetilde{x}_{h^k}$ and $\widetilde{y}_{h^k}$ of the functions $u_{h^k}^f(t)$ and $u_{h^k}^g(t)$, by formula (2.18) we obtain sequentially the approximations $(x_{h^k}(t,h), y_{h^k}(t,h))$ for $k = 1, ..., n$. For $h \in [0, H]$ and $t \in [0, 2\pi + \Delta T]$ we have

$$
\begin{aligned}
x(t,h) &= x_{h^n}(t,h) + \frac{h^{n+1}}{(n+1)!}\frac{\partial^{n+1}x(t,\eta)}{\partial\eta^{n+1}}\Big|_{\eta=h\theta_x(t,h)} = \\
&= x_{h^n}(t,h) + o(h^n) = \sum_{k=1}^n \widetilde{x}_{h^k}(t)h^k + o(h^n), \\
y(t,h) &= y_{h^n}(t,h) + \frac{h^{n+1}}{(n+1)!}\frac{\partial^{n+1}y(t,\eta)}{\partial\eta^{n+1}}\Big|_{\eta=h\theta_y(t,h)} = \\
&= y_{h^n}(t,h) + o(h^n) = \sum_{k=1}^n \widetilde{y}_{h^k}(t)h^k + o(h^n),
\end{aligned}
$$

$$(2.19)$$

$$0 \le \theta_x(t,h) \le 1, \ 0 \le \theta_y(t,h) \le 1.$$

Then by Lemma 2.2 we obtain

$$\widetilde{x}_{h^k}(\cdot), \widetilde{y}_{h^k}(\cdot) \in C^n([0, 2\pi + \Delta T]), \quad k = 1, ..., n. \qquad (2.20)$$

In this case the estimate $o(h^n)$ is uniform $\forall t \in [0, 2\pi + \Delta T]$. From (2.17) and by the choice of initial data in (2.12) we get

$$x_{h^k}(0,h) = x(0,h) = 0, \quad y_{h^k}(0,h) = y(0,h) = h, \qquad k = 1, ..., n.$$

2.2.1.2   *Computation of Lyapunov quantities in the time domain*

Consider the "turn" time $T(h)$ $\big($time of the first crossing the half-line $\{x = 0,\ y > 0\}$ by the solution $(x(t, h), y(t, h))\big)$ for the initial datum $h \in (0, H]$. Complete a definition (by continuity) of the function $T(h)$ at zero: $T(0) = 2\pi$. Since by (2.13) the first approximation of time of crossing the half-line $\{x = 0, y > 0\}$ by solution equals 2π, we can be represented the crossing time as

$$T(h) = 2\pi + \Delta T(h),$$

where $\Delta T(h) = O(h)$. We shall say that $\Delta T(h)$ is a residual of crossing time.

By definition of $T(h)$ we have

$$x(T(h), h) = 0. \tag{2.21}$$

Since by (2.10), $x(\cdot, \cdot)$ has continuous partial derivatives with respect to either arguments up to the order n inc and $\dot{x}(t, h) = \cos(t)h + o(h)$, by the theorem on implicit function the function $T(\cdot)$ is n times differentiable. It is possible to show that $T(h)$ is also differentiable n times at zero (considering, for example, the function $z(t, h) = x(t, h)/h$ and completing its definition at zero by the function $x_{h^1}(t)$ or making use of special theorems of mathematical analysis). By the Taylor formula we obtain

$$T(h) = 2\pi + \sum_{k=1}^{n} \widetilde{T}_k h^k + o(h^n), \tag{2.22}$$

where $\widetilde{T}_k = \dfrac{1}{k!}\dfrac{d^k T(h)}{dh^k}$ (so-called period constants [Gine, 2007]). We shall say that the sum

$$\Delta T_k(h) = \sum_{j=1}^{k} \widetilde{T}_j h^j \tag{2.23}$$

is kth approximation for the residual of the time $T(h)$ of crossing the half-line $\{x = 0,\ y > 0\}$ by solution $\big(x(t, h), y(t, h)\big)$. Substituting relation (2.22) for $t = T(h)$ in the right-hand side of the first equation of (2.19) and denoting the coefficient of h^k by $\widetilde{x}_k$, we obtain the series $x(T(h), h)$ in terms of powers of h:

$$x(T(h), h) = \sum_{k=1}^{n} \widetilde{x}_k h^k + o(h^n). \tag{2.24}$$

In order to express the coefficients $\widetilde{x}_k$ by the coefficients $\widetilde{T}_k$ of the expansion of residual of crossing time we assume that in (2.19) $t = 2\pi + \tau$, then

$$x(2\pi + \tau, h) = \sum_{k=1}^{n} \widetilde{x}_{h^k}(2\pi + \tau)h^k + o(h^n). \qquad (2.25)$$

By smoothness condition (2.20) we have

$$\widetilde{x}_{h^k}(2\pi + \tau) = \widetilde{x}_{h^k}(2\pi) + \sum_{m=1}^{n} \widetilde{x}_{h^k}^{(m)}(2\pi)\frac{\tau^m}{m!} + o(\tau^n), \quad k = 1, ..., n.$$

Substitute this representation in (2.25) for the solution $x(2\pi + \tau, h)$ for $\tau = \Delta T(h)$, and group together the coefficients of the same exponents h. Since $(\Delta T(h))^n = O(h^n)$; then using (2.21) and taking into account (2.22) for $T(h)$, we found

$$h: \quad 0 = \widetilde{x}_1 = \widetilde{x}_{h^1}(2\pi),$$

$$h^2: \quad 0 = \widetilde{x}_2 = \widetilde{x}_{h^2}(2\pi) + \widetilde{x}_{h^1}'(2\pi)\widetilde{T}_1,$$

$$\dots$$

$$h^n: \quad 0 = \widetilde{x}_n = \widetilde{x}_{h^n}(2\pi) + \dots$$

Hence, we sequentially obtain $\widetilde{T}_j$. The coefficients $T_{k=1,...,n-1}$ can be determined sequentially since the expression for $\widetilde{x}_k$ involves only the coefficients $T_{m<k}$ and the factor $\widetilde{x}_{h^1}'(2\pi)$ multiplying T_{k-1} is equal to -1.

We apply a similar procedure to determine the coefficients $\widetilde{y}_k$ of the expansion

$$y(T(h), h) = \sum_{k=1}^{n} \widetilde{y}_k h^k + o(h^n).$$

Substitute the following representation

$$\widetilde{y}_{h^k}(2\pi + \Delta T(h)) = \widetilde{y}_{h^k}(2\pi) + \sum_{m=1}^{n} \widetilde{y}_{h^k}^{(m)}(2\pi)\frac{\Delta T(h)^m}{m!} + o(h^n), \quad k = 1, ..., n$$

in the expression

$$y(2\pi + \Delta T(h), h) = \sum_{k=1}^{n} \widetilde{y}_{h^k}(2\pi + \Delta T(h))h^k + o(h^n).$$

Equating the coefficients of the same exponents h, for the sequential determination of $\widetilde{y}_{i=1,...,n}$ we get

$$h: \quad \widetilde{y}_1 = \widetilde{y}_{h^1}(2\pi),$$

$$h^2: \quad \widetilde{y}_2 = \widetilde{y}_{h^2}(2\pi) + \widetilde{y}_{h^1}'(2\pi)\widetilde{T}_1,$$

$$\dots$$

$$h^n: \quad \widetilde{y}_n = \widetilde{y}_{h^n}(2\pi) + \dots.$$

Here $\widetilde{y}_{h^k=1,..,n}(\cdot)$ and $\widetilde{T}_{k=1,..,n-1}$ are obtained earlier.

Thus, for $n = 2m + 1$ under the condition $f(\cdot,\cdot), g(\cdot,\cdot) \in \mathcal{C}^{(2m+2)}(U)$ we sequentially determined the approximations of the solution $\big(x(t,h), y(t,h)\big)$ at time $t = T(h)$ of first crossing the half-line $\{x = 0, y > 0\}$ accurate to $o(h^{2m+1})$ and the approximation of the time $T(h)$ itself accurate to $o(h^{2m})$. If in this case $\widetilde{y}_k = 0$ for $k = 2,..,2m$, then $\widetilde{y}_{2m+1}$ is mth Lyapunov quantity:

$$L_m = \widetilde{y}_{2m+1}.$$

Realization of this method in Matlab symbolic computation package can be found in [Kuznetsov, 2008].

Example. Consider the Duffing equation represented as the system

$$\dot{x} = -y, \quad \dot{y} = x + x^3. \tag{2.26}$$

For $x_0 = 0, y_0 = h_y$ we have

$$y(t)^2 + x(t)^2 + \frac{1}{2}x(t)^4 = h_y^2. \tag{2.27}$$

For the crossing time $T(h_y)$ from $dt/dy = (x + x^3)^{-1}$ we get

$$
\begin{aligned}
T(h_y) \;=\; & 4 \int_0^{h_y} \frac{dy}{\sqrt{-1 + \sqrt{1 + 2h_y^2 - 2y^2}}\sqrt{1 + 2h_y^2 - 2y^2}} \\[2mm]
=\; & \int_0^{\pi/2} \frac{-h_y \sin(z)\mathrm{d}z}{\sqrt{-1 + \sqrt{1 + 2h_y^2 \sin^2 z}}\sqrt{1 + 2h_y^2 \sin^2 z}} = \\[2mm]
=\; & 2\pi - \frac{3\pi}{4}h_y^2 + \frac{105\pi}{128}h_y^4 - \frac{1155\pi}{1024}h_y^6 + o(h_y^6).
\end{aligned}
$$

Below, we represent a list of the obtained approximations of solution:

$$\widetilde{x}_{h^1}(t) = -\sin(t), \quad \widetilde{y}_{h^1}(t) = \cos(t); \quad \widetilde{x}_{h^2}(t) = \widetilde{y}_{h^2}(t) = 0;$$

$$\widetilde{x}_{h^3}(t) = \frac{1}{8}\cos(t)^2 \sin(t) - \frac{3}{8}t\cos(t) + \frac{1}{4}\sin(t),$$

$$\widetilde{y}_{h^3}(t) = -\frac{3}{8}t\sin(t) + \frac{3}{8}\cos(t) - \frac{3}{8}\cos(t)^3;$$

Here the Lyapunov quantities are equal to zero by virtue of (2.27), and the periodic solution is approximated by a series with nonperiodic coefficients.

2.2.2 Application of Lyapunov function to the computation of Lyapunov quantities

In the case

$$\widetilde{y}_{k=2,\dots,2m} = 0$$

in place of computing of $\widetilde{y}_{2m+1}$ as above one can consider a Lyapunov function and its derivative by virtue of system (2.1):

$$V(x,y) = \frac{(x^2 + y^2)}{2},$$

$$\dot{V}(x,y) = xf(x,y) + yg(x,y).$$

Introduce the notation

$$
\begin{aligned}
L(h) \quad &= \int\limits_{0}^{T(h)} \dot{V}\big(x(t,h), y(t,h)\big)\, \mathrm{d}t \\
&= V\big(x(T(h),h), y(T(h),h)\big) - V\big(x(0,h), y(0,h)\big).
\end{aligned}
$$

The following relation is valid:

Lemma 2.4.

$$
L(h) \quad = \int\limits_{0}^{2\pi + \Delta T_{2m}(h)} \bigg[x_{h^{2m}}(t,h) f\big(x_{h^{2m}}(t,h), y_{h^{2m}}(t,h)\big) + \tag{2.28}
$$

$$
+ y_{h^{2m}}(t,h) g\big(x_{h^{2m}}(t,h), y_{h^{2m}}(t,h)\big) \bigg]\, \mathrm{d}t + o(h^{2m+2}).
$$

Inserting solutions (2.19) into expression (2.28) for $L(h)$, integrating, and grouping together the coefficients of the same powers of h, we get

$$L(h) = \sum_{k=3}^{2m+2} \widetilde{L}_k h^k + o(h^{2m+2}).$$

Here $\widetilde{L}_{3,\dots,2m+1} = 0$ and if $f(\cdot,\cdot), g(\cdot,\cdot) \in \mathcal{C}^{(2m+2)}(U)$ then

$$\widetilde{L}_{2m+2} = \mathrm{L}_m.$$

2.2.3 Lyapunov quantities of Lienard equation

Suppose in (2.1) the following $f(x,y) \equiv 0, g(x,y) = g_{x1}(x)y + g_{x0}(x)$, $g_{x1}(x) = g_{11}x + g_{21}x^2\dots$, $g_{x0}(x) = g_{20}x^2 + g_{30}x^3\dots$. Then we obtain the system (equivalent to Lienard equation)

$$\dot{x} = -y, \quad \dot{y} = x + g_{x1}(x)y + g_{x0}(x). \tag{2.29}$$

Below we consider and use the expressions for Lyapunov quantities $L_{i=1,\dots,5}$ for Lienard equation, which are computed in the work [Leonov and Kuznetsova, 2009]. For the first Lyapunov quantity we have

$$L_1 = -\frac{\pi}{4}(g_{20}g_{11} - g_{21}).$$

If $g_{21} = g_{20}g_{11}$, then we get $L_1 = 0$ and

$$L_2 = \frac{\pi}{24}(3g_{41} - 5g_{20}g_{31} - 3g_{40}g_{11} + 5g_{20}g_{30}g_{11}).$$

If $g_{41} = \frac{5}{3}g_{20}g_{31} + g_{40}g_{11} - \frac{5}{3}g_{20}g_{30}g_{11}$, then we get $L_2 = 0$ and

$$L_3 = -\frac{\pi}{576}(70g_{20}^3g_{30}g_{11} + 105g_{20}g_{51} + 105g_{30}^2g_{11}g_{20} + 63g_{40}g_{31} - 63g_{11}g_{40}g_{30} - 105g_{30}g_{31}g_{20} - 70g_{20}^3g_{31} - 45g_{61} - 105g_{50}g_{11}g_{20} + 45g_{60}g_{11}).$$

If g_{61} is determined from equation $L_3 = 0$, then we get

$$L_4 = \frac{1}{17280}(945g_{81} + 4158g_{20}^2g_{40}g_{31} + 2835g_{20}g_{30}g_{51} - 5670g_{20}g_{30}g_{11}g_{50} - 4158g_{20}^2g_{30}g_{11}g_{40} + 2835g_{20}g_{11}g_{70} + 1215g_{30}g_{11}g_{60} + 1701g_{40}g_{11}g_{50} - 4620g_{20}^3g_{11}g_{50} - 8820g_{20}^3g_{30}g_{31} + 1701g_{30}g_{40}g_{31} + 2835g_{20}g_{50}g_{31} - 2835g_{20}g_{30}^2g_{31} - 1701g_{30}^2g_{11}g_{40} + 8820g_{20}^3g_{30}^2g_{11} + 3080g_{20}^5g_{11}g_{30} + 2835g_{20}g_{30}^3g_{11} + 4620g_{20}^3g_{51} - 1701g_{40}g_{51} - 945g_{11}g_{80} - 3080g_{20}^5g_{31} - 1215g_{60}g_{31} - 2835g_{20}g_{71}).$$

If g_{81} is determined from equation $L_4 = 0$ then we get

$$L_5 = \frac{\pi}{3110400}(-1621620g_{20}^2g_{40}g_{11}g_{50} - 3118500g_{20}^2g_{30}g_{40}g_{31} - 935550g_{20}g_{30}g_{11}g_{70} + 2522520g_{20}^4g_{30}g_{11}g_{40} - 935550g_{20}g_{30}g_{50}g_{31} - 486486g_{20}g_{30}g_{40}^2g_{11} + 1403325g_{20}g_{30}^2g_{11}g_{50} - 579150g_{20}^2g_{30}g_{11}g_{60} + 5128200g_{20}^3g_{30}g_{11}g_{50} - 561330g_{30}g_{40}g_{11}g_{50} + 127575g_{101} + 1351350g_{20}^3g_{71} - 127575g_{11}g_{100} - 280665g_{40}g_{71} - 200475g_{60}g_{51} - 2402400g_{20}^5g_{51} + 1601600g_{20}^7g_{31} - 155925g_{80}g_{31} - 1601600g_{20}^7g_{11}g_{30} + 2402400g_{20}^5g_{11}g_{50} - 467775g_{20}g_{50}^2g_{11} - 4158000g_{20}^3g_{30}^3g_{11} + 1621620g_{20}^2g_{40}g_{51} + 467775g_{20}g_{70}g_{31} - 467775g_{20}g_{30}^2g_{51} + 467775g_{20}g_{30}^3g_{31} + 200475g_{30}g_{60}g_{31} + 5613300g_{20}^5g_{30}g_{31} + 467775g_{20}g_{30}g_{71} - 2113650g_{20}^3g_{50}g_{31} + 280665g_{40}g_{50}g_{31} - 2522520g_{20}^4g_{40}g_{31} - 280665g_{30}^2g_{40}g_{31} + 486486g_{20}g_{40}^2g_{31} - 5613300g_{20}^5g_{30}g_{11} + 467775g_{20}g_{11}g_{90} + 280665g_{30}g_{11}g_{40} - 467775g_{20}g_{30}^4g_{11} + 467775g_{20}g_{50}g_{51} + 280665g_{30}g_{40}g_{51} - 3014550g_{20}^2g_{30}g_{51} + 4158000g_{20}^3g_{30}^2g_{31} + 579150g_{20}^2g_{60}g_{31} - 1351350g_{20}^3g_{11}g_{70} + 280665g_{40}g_{11}g_{70} + 200475g_{50}g_{11}g_{60} - 200475g_{30}^2g_{11}g_{60} + 155925g_{30}g_{11}g_{80} + 3118500g_{20}^2g_{30}^2g_{11}g_{40} - 467775g_{20}g_{91}).$$

If $L_{1,\dots,n-1} = 0$ and $L_n \neq 0$, then using the famous technique of Bautin we can construct here n small limit cycles by small disturbances of system coefficients [Bautin, 1952; Lloyd and Pearson, 1997; Lynch, 2005].

2.3 Transformation between quadratic systems and Lienard equation

Consider a special class [Cherkas, 1976; Leonov, 1997; Leonov, 1998] of Lienard equations

$$\begin{aligned} \dot{x} &= y, \\ \dot{y} &= -f(x)y - g(x), \end{aligned} \tag{2.30}$$

which is obtained from a nontrivial quadratic system with zero equilibrium

$$\begin{aligned} \dot{x} &= p(x,y) = a_1 x^2 + b_1 xy + c_1 x^2 + \alpha_1 x + \beta_1 y, \\ \dot{y} &= q(x,y) = a_2 x^2 + b_2 xy + c_2 y^2 + \alpha_2 x + \beta_2 y, \end{aligned} \tag{2.31}$$

where $c_1 = 0$ (the latter always can be obtained by the change $x \to x + \nu y$ in the case $a_1 \neq 0$ or the reassignment $x \leftrightarrow y$ in the case $a_1 = 0$) and $b_1 \beta_1 \neq 0$.

Suppose,

$$\begin{aligned} f(x) &= (Ax + B))|x + 1|^{q-2}, \\ g(x) &= (C_1 x^3 + C_2 x^2 + C_3 x + 1)x \frac{|x+1|^{2q}}{(x+1)^3}. \end{aligned} \tag{2.32}$$

Consider the nondegenerate case

$$A \neq B, \ AB \neq 0, \ q \neq \frac{1}{2}.$$

Then the following assertion is valid

Lemma 2.5. *Suppose, for the coefficients A, B, C_1, C_2, C_3, q of system (2.30) the relations*

$$\begin{aligned} \frac{(B-A)}{(2q-1)^2}((1-q)B + (3q-2)A) &= 2C_2 - 3C_1 - C_3, \\ \frac{(B-A)}{(2q-1)^2}(B + 2(q-1)A) &= C_2 - 2C_1 - 1. \end{aligned}$$

are satisfied. Then equation (2.30) can be reduced to quadratic system (2.31) with the coefficients

$$b_1 = 1, c_1 = 0, \alpha_1 = 1, \beta_1 = 1, c_2 = -q, \alpha_2 = -2, \beta_2 = -1,$$

$$a_1 = 1 + \frac{B-A}{2q-1}, \ a_2 = -(q+1)a_1^2 - Aa_1 - C_1, \ b_2 = -A - a_1(2q+1).$$

In this case for the Lyapunov quantities we have the following

Lemma 2.6. *if $L_1 = 0$ and*

$$A = \frac{2B(q+2)}{5}$$

then $L_2 = 0$ and

$$C_1 = (q+3)\frac{B^2}{25} - \frac{(1+3q)}{5},$$

$$C_2 = \left(15(1-2q) + 3B^2\right)\frac{1}{25}, \quad C_3 = \frac{3(3-q)}{5} \tag{2.33}$$

$$L_3 = -\frac{\pi B(q+2)(3q+1)[5(q+1)(2q-1)^2 + B^2(q-3)]}{20000}. \tag{2.34}$$

Thus, if the conditions of Lemma 2.6 and relation $L_3 \neq 0$ are valid, then by small disturbances of system we can obtain three small limit cycles for quadratic system around the zero equilibrium.

Note, that here $L_3 = 0$ implies $L_4 = 0$.

Lemma 2.7. *For $c_1 = 0, b_1 \neq 0$, system (2.31) can be reduced to Lienard equation (2.30) with the functions*

$$F(x) = R(x)e^{p(x)} = R(x)|\beta_1 + b_1 x|^q,$$

$$G(x) = P(x)e^{2p(x)} = P(x)|\beta_1 + b_1 x|^{2q}.$$

Here $q = -\dfrac{c_2}{b_1}$,

$$R(x) = -\frac{(b_1 b_2 - 2a_1 c_2 + a_1 b_1)x^2 + (b_2\beta_1 + b_1\beta_2 - 2\alpha_1 c_2 + 2a_1\beta_1)x}{(\beta_1 + b_1 x)^2} +$$

$$+ \frac{\alpha_1\beta_1 + \beta_1\beta_2}{(\beta_1 + b_1 x)^2},$$

$$P(x) = -\left(\frac{a_2 x^2 + \alpha_2 x}{\beta_1 + b_1 x} - \frac{(b_2 x + \beta_2)(a_1 x^2 + \alpha_1 x)}{(\beta_1 + b_1 x)^2} + \frac{c_2(a_1 x^2 + \alpha_1 x)^2}{(\beta_1 + b_1 x)^3}\right).$$

2.4 Method of asymptotic integration for Lienard equation with discontinuous right–hand side and large limit cycles

In [Leonov, 2009abc] new method of asymptotic integration for Lienard equation is suggested which is effective for attractors localization and investigation of limit cycles.

Consider Lienard equation (2.30):

$$\dot{x} = y, \ \dot{y} = -f(x)y - g(x)$$

$$f(x) = \phi(x)|x+1|^{q-2}, \quad g(x) = \psi(x)\frac{|x+1|^{2q}}{(x+1)^3}, \tag{2.35}$$

where $\phi(x)$ and $\psi(x)$ are smooth functions on $(-\infty, +\infty)$, $\psi(0) = 0, q \in (-1, 1)$. Suppose,

$$f(x) = (A + O(\frac{1}{|x|}))|x|^q, \ g(x) = (C + O(\frac{1}{|x|}))|x|^q \tag{2.36}$$

Hold some numbers $a > -1$ and $\delta > 0$ fixed, and consider sufficiently large numbers R and $\widetilde{R}$. Then the following results occur

Lemma 2.8. *Let* $C > 0, 4C(q+1) > A^2$. *Then for the solution of system (2.35) with the initial data* $x(0) = a$, $y(0) = R$ *there exists a number* $T > 0$ *such that*

$$x(T) = a, \ y(T) < 0, x(t) > a, \ \forall t \in (0, T),$$

$$R\exp\left(\frac{\lambda\pi}{\omega} - \delta\right) < |y(T)| < R\exp\left(\frac{\lambda\pi}{\omega} + \delta\right).$$

Here

$$\omega = \frac{\sqrt{4(q+1)C - A^2}}{2(q+1)}, \lambda = \frac{-A}{2(q+1)}.$$

Consider the behavior of system near $x = -1$. Suppose,

$$f(x) = (P + O(|x+1|))|x+1|^{q-2}, \ g(x) = (Q + O(|x+1|))\frac{|x+1|^{2q}}{(x+1)^3} \tag{2.37}$$

Lemma 2.9. *If* $P > 0, P^2 > 4Q(q-1) > 0$, *then for the solution of system (2.35) with the initial data* $x(0) = a$, $y(0) = -\widetilde{R}$ *there exists a number* $T > 0$ *such that*

$$x(T) = a, \ 0 < y(T) < \delta\widetilde{R}, x(t) \in (-1, a), \ \forall t \in (0, T).$$

These lemmas imply the following

Theorem 2.1. *We assume that the function* $g(x)$ *has the unique zero at* $x = x_1$ *in the interval* $(-1, +\infty)$ *and the equilibrium* $(x = x_1, y = 0$ *is locally unstable. Then system (2.35) has limit cycles in the domain* $x \in (-1, +\infty), y \in \mathbb{R}^1$.

A similar assertions are also valid for the left half-plane.

2.5 Four limit cycles for Lienard equation and the corresponding quadratic system

Suppose, the conditions of Lemma 2.6 and relation $L_3 \neq 0$ are satisfied, then by small disturbances of coefficients we can obtain three small limit cycles around zero equilibrium of system and seek large limit cycles on a plane of the rest two coefficients (B, q).

By Theorem 2.1 for Lienard system (and for the corresponding to it quadratic system) it can be obtained analytically the estimates of the domain of parameters (B, q), corresponding to the existence of large limit cycle to the left of the line of discontinuity $x = -1$. Further, for study of this domain we apply the method of asymptotical integration for which purpose we make the change of variables $x = -v - 2$ (symmetric mapping with respect to the line of discontinuity $x = -1$) and, group together the coefficients of the same powers of $(1 + v)$ in the Lienard equation. In this case we obtain

$$\ddot{v} + f(v)\dot{v} + g(v) = 0, \qquad (2.38)$$

$$f(v) = \big(A_v(1 + v)^2 + B_v(v + 1) + C_v\big)|v + 1|^{q-2},$$

$$g(v) = \bigg(C_{v1}(1 + v)^4 + C_{v2}(1 + v)^3 + C_{v3}(1 + v)^2 +$$

$$+ C_{v4}(1 + v) + C_{v5}\bigg)\frac{|v + 1|^{2q}}{(v + 1)^3}. \qquad (2.39)$$

By conditions (2.33) we have

$$A_v = \frac{2}{5}B(2 + q), \ B_v = \frac{1}{5}B(3 + 4q), \ C_v = \frac{1}{5}B(-1 + 2q),$$

$$C_{v1} = \frac{1}{25}\big((3 + q)B^2 - 15q - 5\big), \ C_{v2} = \frac{1}{25}\big((4q + 9)B^2 - 35 - 30q\big),$$

$$C_{v3} = \frac{1}{25}\big((6q + 9)B^2 - 30 - 15q\big), \ C_{v4} = \frac{1}{25}B^2(3 + 4q), \ C_{v5} = \frac{1}{25}B^2q.$$

Now we assume here that,

$$B < 0, \ q \in (-1, 0).$$

Then to the right of the line of discontinuity $(v = -1)$ system (2.38) has only one equilibrium and the method of asymptotic integration considered above makes it possible to obtain existence conditions for large limit cycle.

We consider the special changes $v = D(z)$ for reducing system (2.38) to the system $\ddot{z} + \widetilde{f}(z)\dot{z} + \widetilde{g}(z) = 0$ with the functions f and g of the form (2.36) and (2.36), respectively, namely

1) for $v \in (-1, +\infty)$ by the change $v = D(z) = z^{\frac{1}{q+1}} - 1$ we obtain

$$\widetilde{f}(z) = \frac{1}{(q+1)}(A_v + B_v z^{-\frac{1}{q+1}} + C_v z^{-\frac{2}{q+1}}),$$

$$\widetilde{g}(z) = \frac{1}{(q+1)}z(C_{v1} + z^{-\frac{1}{q+1}}C_{v2} + z^{-\frac{2}{q+1}}C_{v3} + z^{-\frac{3}{q+1}}C_{v4} + z^{-\frac{4}{q+1}}C_{v5}).$$

2) for $v \in (-1, 0)$ by the change $v = D(z) = z^{\frac{1}{q-1}} - 1$ we obtain

$$\widetilde{f}(z) = \frac{1}{(q-1)}(A_v z^{\frac{2}{q-1}} + B_v z^{\frac{1}{q-1}} + C_v),$$

$$\widetilde{g}(z) = \frac{1}{(q-1)}z(z^{\frac{4}{q-1}}C_{v1} + z^{\frac{3}{q-1}}C_{v2} + z^{\frac{3}{q-1}}C_{v3} + z^{\frac{4}{q-1}}C_{v4} + C_{v5}).$$

Applying to the obtained equation the method of asymptotic integration and taking into account the symmetry with respect to B in (2.34) and (2.38), we obtain [Leonov, 2009a]

Theorem. *Let*

$$B^2 < -5(q+1)(3q+1), \quad B \neq 0 \qquad (2.40)$$

be valid. Then system (2.30) with functions (2.32) has a limit cycle in the half-plane $\{x < -1, \quad y \in R^1\}$.

In the figure (below), on the plane of parameters B, q the domain Ω, where (2.40) is satisfied, is discriminated. The interior of domain, corresponding to the existence of four cycles has been obtained by Shi [Shi, 1980] for quadratic systems. Noted that this domain lies entirely in Ω.

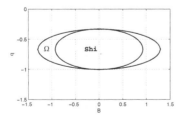

Thus, we generalize the well-known theorem of Shi on the existence of four cycles for quadratic systems.

Acknowledgments

This work partly supported by the Dutch-Russian scientific cooperation programme 047.017.018, projects of Ministry of education and science of RF, Grants board of President RF, RFBR and CONACYT.

References

Andronov, A. A., Vitt, E. A. and Khaiken S. E. (1966). *Theory of Oscillators* Pergamon Press. Oxford.

Anosov, D. V., Aranson, S. Kh., Arnold, V. I., Bronshtein, I. U., Grines, V. Z. and Il'yashenko, Yu. S. (1997). *Ordinary Differential Equations and Smooth Dynamical Systems*. Springer. New York.

Arnold, V. I. (2005). *Experimental Mathematics [in Russian]*. Fazis. Moscow.

Bautin, N. (1939). Du nombre de cycles limites naissant en cas de variation des coefficients dunetat dequilibre du type foyer ou centre [in Russian]. *C. R. (Doklady) Acad. Sci. URSS (N. S.)* **24**, pp. 669–672.

Bautin, N. (1952). On the number of limit cycles arising with the variation of the coefficients from an equilibrium point of focus or center type [in Russian]. *Mat. Sbornik*, **30**(72), pp. 181–196. [Transl. into English (1954) in reprint, *AMS Transl.* (1) 5 (1962), pp. 396–413]

Bautin, N. N., Leontovich, E. A. (1976). *Methods and procedures for qualitative study of dynamical systems on plane [in Russian]*. Nauka. Moscow.

Cesari, L. (1959). *Asymptotic Behavior and Stability Problems in Ordinary Differential Equations*. Springer. Berlin.

Chavarriga, J. and Grau, M. (2003). Some open problems related to 16th Hilbert problem, *Sci. Ser. A Math. Sci. (N.S.)*, 9, pp. 1–26.

Cherkas, L. A. (1976). Number of limit cycles of an autonomous second-order system. *Differential Equations*, 5, pp. 666–668.

Christopher, C. and Li, Ch. (2007). *Limit cycles of differential equations. Advanced Courses in Mathematics* CRM Barcelona. Basel. Birkhauser Verlag.

Dumortier, F., Llibre, J. and Artes, J. (2006). *Qualitative Theory of Planar Differential Systems*. Springer.

Gasull, A., Guillamon, A. and Manosa, V. (1997). An explicit expression of the first Liapunov and period constants with applications, *J. Math. Anal. Appl.*, 211, pp. 190–212.

Gasull, A., Guillamon, A., and Manosa, V. (1999). An Analytic-Numerical Method for Computation of the Liapunov and Period Constants Derived from Their Algebraic Structure. *SIAM J. Numer. Anal* **36**(4), pp. 1030–1043.

Gine, J. (2007). On some problems in planar differential systems and Hilbert's 16th problem, *Chaos, Solutions and Fractals*, **31**, pp. 1118–1134.

Hartman, P. (1964). *Ordinary differential equation*. John Willey & Sons.

New York.

Kuznetsov, N. V. (2008). *Stability and Oscillations of Dynamical Systems: Theory and Applications.* Jyväskylä University Printing House.

Kuznetsov, N. V. and Leonov, G. A. (2008). Computation of Lyapunov quantities. *6th EUROMECH Nonlinear Dynamics Conference* (http://lib.physcon.ru/?item=1802)

Kuznetsov, N. V. and Leonov, G. A. (2008). Limit cycles and strange behavior of trajectories in two dimension quadratic systems, *Journal of Vibroengineering,* **10**(4), pp. 460–467.

Lefschetz, S. (1957). *Differential Equations: Geometric Theory.* Interscience Publishers. New York.

Leonov, G. A. (1997). Two-Dimensional Quadratic Systems as a Lienard Equation. *Differential Equations and Dynamical Systems,* **5**(3/4), pp. 289–297.

Leonov, G. A. (1998). The problem of estimation of the number of cycles of two-dimensional quadratic systems from nonlinear mechanics point of view. *Ukr. Math. J.,* **50**(1), pp. 48–57.

Leonov, G. A. and Kuznetsov, N. V. (2007). Time-Varying Linearization and the Perron effects. *International journal of bifurcation and chaos,* **17**(4), pp. 1–29.

Leonov, G. A., Kuznetsov, N. V. and Kudryashova, E. V. (2008). Limit cycles of two dimensional systems. Calculations, proofs, experiments. *Vestnik St.Petersburg University. Mathematics,* **41**(3), pp. 216–250.

Leonov, G. A. (2009). Limit cycles of Lienard system with discontinuous coefficients. *Doclady Akademii Nauk,* **426**(1), pp. 47–50.

Leonov, G. A. (2009). Effective methods for investigation of limit cycles in dynamical systems (2009) *Applied Mathematics and Mechanics,* 4, [in print]

Leonov, G. A. (2009). The criteria of four cycles existence in quadratic systems. *Applied Mathematics and Mechanics,* 5, [in print]

Leonov, G. A. and Kuznetsova O. A. (2009). Evaluation of the First Five Lyapunov Exponents for the Lienard System. *Doklady Physics,* **543**, pp. 131–133.

Lloyd, N. G. (1988). Limit cycles of polynomial systems – some recent developments. In *New Direction in Dynamical Systems.* Cambridge University Press, pp. 192–234.

Lloyd, N. G. and Pearson, J. (1997). Five limit cycles for a simple cubic system. *Publicacions Matem'atiques,* **41**, pp. 199–208.

Lyapunov, A. M. (1892). *The General Problem of the Stability of Motion [in Russian].* Kharkov. [Transl. into English (1966) *Stability of Motion.* New York and London: Academic Press.]

Lynch, S. (2005). Symbolic computation of Lyapunov quantities and the second part of Hilbert's sixteenth problem. *Differential Equations with Symbolic Computations,* Wang, Dongming; Zheng, Zhiming (Eds.), Series: Trends in Mathematics, pp. 1–26 ISBN: 3-7643-7368-7

Marsden, J. and McCracken, M. (1976). *Hopf bifurcation and its applications.* Springer. New York.

Reyn, J. W. (1994). *A bibliography of the qualitative theory of quadratic systems of differential equations in the plane. Third edition. Report 94-02.* Delft University of Technology.

Roussarie, R. (1998). *Bifurcations of planar vector fields and Hilbert's sixteenth problem. Progress in mathematics. V.164.* Birkhauser. Basel-Boston-Berlin.

Sabatini, M. and Chavarriga, J. (1999). A survey of isochronous centers. *Qualitative Theory of Dynamical Systems*, **1**, pp. 1–70.

Serebryakova, N. (1959). Behavior of dynamical system with one degree of freedom near the points of boundaries of domain of stability in the case when a safe boundary goes to the dangerous one [in Russian]. *Izvestiya AN SSSR. Division of technical sciences, mechanics and engineering*, **2**, pp. 178–182.

Shi, S. L. (1980). A concrete example of the existence of four limit cycles for plane quadratic systems. *Sci. Sinica*, 23, pp. 153–158.

Shilnikov, L., Turaev, D. and Chua, L. (2001). *Methods of Qualitative Theory in Nonlinear Dynamics: Part 2.* World Scientific.

Yu, P. (1998). Computation of normal forms via a perturbation technique. *J. Sound Vibr.*, **211**, pp. 19–38.

Yu, P. and Chen, G. (2008). Computation of focus values with applications. *Nonlinear Dynamics*, **51**(3), pp. 409–427.

Absolute Observation Stability for Evolutionary Variational Inequalities

G.A. Leonov, V. Reitman

Department of Mathematics and Mechanics
St. Petersburg State University
Russia

Abstract

We derive absolute observation stability and instability results for controlled evolutionary inequalities which are based on frequency-domain characteristics of the linear part of the inequalities. The uncertainty parts of the inequalities (nonlinearities which represent external forces and constitutive laws) are described by certain local and integral quadratic constraints. Other terms in the considered evolutionary inequalities represent contact-type properties of a mechanical system with dry friction.

3.1 Introduction

Suppose that Y_0 is a real Hilbert space. We denote by $(\cdot,\cdot)_0$ and $\|\cdot\|_0$ the scalar product resp. the norm on Y_0. Let $A : \mathcal{D}(A) \to Y_0$ be the generator of a C_0-semigroup on Y_0 and define the set $Y_1 := \mathcal{D}(A)$. Here $\mathcal{D}(A)$ is the domain of A, which is dense in Y_0 since A is a generator. We denote with $\rho(A)$ the resolvent set of A. The spectrum of A, which is the complement of $\rho(A)$, is denoted by $\sigma(A)$. If we define with an arbitrary but fixed $\beta \in \rho(A) \cap \mathbb{R}$ for any $y, \eta \in Y_1$, the value

$$(y, \eta)_1 := ((\beta I - A)y, \, (\beta I - A)\eta)_0 \, , \tag{3.1}$$

then the set Y_1 equipped with this scalar product $(\cdot, \cdot)_1$ and the corresponding norm $\| \cdot \|_1$ becomes a Hilbert space (different numbers β give different but equivalent norms). Denote by Y_{-1} the Hilbert space which is the completion of Y_0 with respect to the norm $\|y\|_{-1} := \|(\beta I - A)^{-1}y\|_0$ and which has the corresponding scalar product

$$(y, \eta)_{-1} := \left((\beta I - A)^{-1}y, \, (\beta I - A)^{-1}\eta \right)_0 ,$$

$$\forall \, y, \eta \in Y_{-1}. \tag{3.2}$$

Thus, we get the inclusions $Y_1 \subset Y_0 \subset Y_{-1}$, which are dense with continuous embedding, i.e. for $\alpha = 1, 0$ $Y_\alpha \subset Y_{\alpha-1}$, is dense and $\|y\|_{\alpha-1} \leq c\|y\|_\alpha$, $\forall \, y \in Y_\alpha$. Sometimes [Banks et al., 1997, Banks and Ito, 1988] the introduced triple of spaces (Y_1, Y_0, Y_{-1}) is called a *Gelfand triple*. The pair (Y_1, Y_{-1}) is also called *Hilbert rigging* of the *pivot space* Y_0, Y_1 is an *interpolation space* of Y_0, and Y_{-1} is an *extrapolation space* of Y_0. Since for any $y \in Y_0$ and $z \in Y_1$ we have

$$|(y, z)_0| = |(\beta I - A)^{-1}y, ((\beta I - A)z)_0|$$

$$\leq \|y\|_{-1}\|z\|_1 \, , \tag{3.3}$$

we can extend $(\cdot, z)_0$ by continuity onto Y_{-1} obtaining the inequality

$$|(y, z)_0| \leq \|y\|_{-1}\|z\|_1 \, , \quad \forall \, y \in Y_{-1}, \forall \, z \in Y_1.$$

Let us denote this extension also by $(\cdot, \cdot)_{-1,1}$ and call it *duality pairing* on $Y_{-1} \times Y_1$. The operator A has a unique extension to an operator in $\mathcal{L}(Y_0, Y_{-1})$ which we denote by the same symbol. Suppose now that $T > 0$ is arbitrary and define the norm for Bochner measurable functions in $L^2(0, T; Y_j)$ $(j = 1, 0, -1)$ through

$$\|y(\cdot)\|_{2,j} := \left(\int\limits_0^T \|y(t)\|_j^2 \, dt \right)^{1/2}. \tag{3.4}$$

Let $\mathcal{L}_T$ be the space of functions such that $y \in L^2(0, T; Y_1)$ and $\dot{y} \in L^2(0, T; Y_{-1})$, where the time derivative $\dot{y}$ is understood in the sense of distributions with values in a Hilbert space. The space $\mathcal{L}_T$ equipped with the norm

$$\|y\|_{\mathcal{L}_T} := \left(\|y(\cdot)\|_{2,1}^2 + \|\dot{y}(\cdot)\|_{2,-1}^2 \right)^{1/2} \tag{3.5}$$

is a Hilbert space and will be used for the description of solutions to evolutionary systems.

---

[1]Supported by DAAD (German Academic Exchange Service)

3.2   Evolutionary variational inequalities

Suppose that $T > 0$ is arbitrary and consider for a.a. $t \in [0, T]$ the observed and controlled evolutionary variational inequality

$$(\dot{y} - Ay - B\xi - f(t), \eta - y)_{-1,1} \tag{3.6}$$

$$+ \psi(\eta) - \psi(y) \geq 0, \qquad \forall \, \eta \in Y_1$$

$$y(0) = y_0 \in Y_0 \,,$$

$$w(t) = Cy(t) \,, \quad \xi(t) \in \varphi(t, w(t)) \,, \tag{3.7}$$

$$\xi(0) = \xi_0 \in \mathcal{E}(y_0) \,,$$

$$z(t) = Dy(t) + E\,\xi(t) \,. \tag{3.8}$$

In equations (3.6) – (3.8) it is supposed that $C \in \mathcal{L}(Y_{-1}, W), D \in \mathcal{L}(Y_1, Z)$ and $E \in \mathcal{L}(\Xi, Z)$ are linear operators, Ξ, W and Z are real Hilbert spaces, $Y_1 \subset Y_0 \subset Y_{-1}$ is a real Gelfand triple and $A \in \mathcal{L}(Y_0, Y_{-1}), B \in \mathcal{L}(\Xi, Y_{-1}), \varphi : \mathbb{R}_+ \times W \to 2^{\Xi}$ is a set-valued map, $\psi : Y_1 \to \mathbb{R}_+$ and $f : \mathbb{R}_+ \to Y_{-1}$ are given nonlinear maps. The calculation of $\xi(t)$ in (3.7) shows that this value in general also depends on certain "initial state" ξ_0 of φ taken from a set $\mathcal{E}(y_0) \subset \Xi$. This situation is typical for hysteresis nonlinearities [Reitmann, 2005].

In the following we denote by $\| \cdot \|_{\Xi}$, $\| \cdot \|_W$ and $\| \cdot \|_Z$ the norm in Ξ, W resp. Z.

Let us now introduce the solution space for the problem (3.6), (3.7).

Definition 3.1. Any pair of functions $\{y(\cdot), \xi(\cdot)\}$ with $y \in \mathcal{L}_T$ and $\xi \in L^2_{\text{loc}}(0, \infty; \Xi)$ such that $B\xi \in \mathcal{L}_T$, satisfying (2.1), (2.2) almost everywhere on $(0, T)$, is called **solution of the Cauchy problem** $y(0) = y_0$, $\xi(0) = \xi_0$ defined for (2.1), (2.2).

In order to have an existence property for (2.1), (2.2) we state the following assumption:

(C1) The Cauchy-problem (2.1), (2.2) has for arbitrary $y_0 \in Y_0$ and $\xi_0 \in \mathcal{E}(y_0) \subset \Xi$ at least one solution $\{y(\cdot), \xi(\cdot)\}$.

Assumption **(C1)** is fulfilled, for example, in the following situation [Pankov, 1986].

(C2) a) The nonlinearity $\varphi : \mathbb{R}_+ \times W \to \Xi$ is a function having the property that $\mathcal{A}(t) := -A - B\varphi(t, C\cdot) : Y_1 \to Y_{-1}$ is a family of monotone hemicontinuous operators such that the inequality

$$\|\mathcal{A}(t)y\|_{-1} \leq c_1 \|y\|_1 + c_2 \,, \quad \forall \, y \in Y_1 \,,$$

is satisfied, where $c_1 > 0$ and $c_2 \in \mathbb{R}$ are constants not depending on $t \in [0, T]$. Furthermore for any $y \in Y_1$ and for any bounded set $U \subset Y_1$ the family of functions $\{(\mathcal{A}(t)\eta, y)_{-1,1}, \eta \in U\}$ is equicontinuous with respect to t on any compact subinterval of $\mathbb{R}_+$.

b) ψ is a proper, convex, and semicontinuous from below function on $\mathcal{D}(\psi) \subset Y_1$.

(C3) $f \in L^2_{\text{loc}}(\mathbb{R}_+; Y_{-1})$.

(C4) In the sequel we consider only solutions y of (3.6),(3.7) for which $\dot{y}$ belongs to $L^2_{\text{loc}}(\mathbb{R}; Y_{-1})$.

a) Note that in the special case when $\psi \equiv 0$ in (3.6) the evolutionary variational inequality is equivalent for a.a. $t \in [0, T]$ to the equation

$$\dot{y} = Ay + B\xi + f(t) \quad \text{in } Y_{-1},$$
$$y(0) = Y_0, \ w(t) = Cy(t), \ \xi(t) \in \varphi(t, w(t)),$$
$$\xi(0) \in \mathcal{E}(y_0),$$
$$z(t) = Dy(t) + E\xi(t).$$

Under the assumption that φ is a single valued function this class was considered in [Banks *et al.*, 1997; Banks and Ito, 1988; Brusin, 1976] .

Definition 3.2. a) Suppose F and G are quadratic forms on $Y_1 \times \Xi$. The **class of nonlinearities** $\mathcal{N}(F, G)$ defined by F and G consists of all maps $\varphi : \mathbb{R}_+ \times W \to 2^\Xi$ such that for any $y(\cdot) \in L^2_{\text{loc}}(0, \infty; Y_1)$ with $\dot{y}(\cdot) \in L^2_{\text{loc}}(0, \infty; Y_{-1})$ and any $\xi(\cdot) \in L^2_{\text{loc}}(0, \infty; \Xi)$ with $\xi(t) \in \varphi(t, Cy(t))$ for a.e. $t \geq 0$, it follows that $F(y(t), \xi(t)) \geq 0$ for a.e. $t \geq 0$ and (for any such pair $\{y, \xi\}$) there exists a continuous functional $\Phi : W \to \mathbb{R}$ such that for any times $0 \leq s < t$ we have $\displaystyle\int_s^t G(y(\tau), \xi(\tau))d\tau \geq \Phi(Cy(t)) - \Phi(Cy(s))$.

b) The **class of functionals** $\mathcal{M}(d)$ defined by a constant $d > 0$ consists of all maps $\psi : Y_1 \to \mathbb{R}_+$ such that for any $y \in L^2_{\text{loc}}(0, \infty; Y_0)$ with $\dot{y} \in L^2_{\text{loc}}(0, \infty; Y_1)$ the function $t \mapsto \psi(y(t))$ belongs to $L^1(0, \infty; \mathbb{R})$ satisfying $\displaystyle\int_0^\infty \psi(y(t))dt \leq d$ and for any $\varphi \in \mathcal{N}(F, G)$ and any $\psi \in \mathcal{M}(d)$ the Cauchy-problem (3.6) $-$ (3.8) has a solution $\{y(\cdot), \xi(\cdot)\}$ on any time interval $[0, T]$.

3.3 Basic assumptions

(F1) The operator $A \in \mathcal{L}(Y_1, Y_{-1})$ is *regular* [Duvant and Lions, 1976; Likhtarnikov and Yakubovich, 1976] , i.e., for any $T > 0, y_0 \in Y_1, \psi_T \in Y_1$ and $f \in L^2(0, T; Y_0)$ the solutions of the direct problem

$$\dot{y} = Ay + f(t), \; y(0) = y_0, \quad \text{a.a. } t \in [0, T]$$

and of the dual problem

$$\dot{\psi} = -A^*\psi + f(t), \; \psi(T) = \psi_T, \quad \text{a.a. } t \in [0, T]$$

are strongly continuous in t in the norm of Y_1. Here (and in the following) $A^* \in \mathcal{L}(Y_{-1}, Y_0)$ denotes the adjoint to A, i.e., $(Ay, \eta)_{-1,1} = (y, A^*\eta)_{-1,1}, \; \forall \, y, \eta \in Y_1$.

The assumption **(F1)** is satisfied [Likhtarnikov and Yakubovich, 1976] if the embedding $Y_1 \subset Y_0$ is completely continuous, i.e., transforms bounded sets from Y_1 into compact sets in Y_0.

(F2) The pair (A, B) is L^2-*controllable*, [Brusin, 1976; Likhtarnikov and Yakubovich, 1976] i.e., for arbitrary $y_0 \in Y_0$ there exists a control $\xi(\cdot) \in L^2(0, \infty; \Xi)$ such that the problem

$$\dot{y} = Ay + B\xi, \quad y(0) = y_0$$

is well-posed on the semiaxis $[0, +\infty)$, i.e., there exists a solution $y(\cdot) \in \mathcal{L}_\infty$ with $y(0) = y_0$.

It is easy to see that a pair (A, B) is L^2-controllable if this pair is *exponentially stabilizable*, i.e., if an operator $K \in \mathcal{L}(Y_0, \Xi)$ exists such that the solution $y(\cdot)$ of the Cauchy-problem $\dot{y} = (A + BK)y, \; y(0) = y_0$, decreases exponentially as $t \to \infty$, i.e.,

$$\exists \, c > 0 \quad \exists \, \varepsilon > 0 : \|y(t)\|_0 \leq c \, e^{-\varepsilon t} \|y_0\|_0, \; \forall \, t \geq 0.$$

(F3) Let $F(y, \xi)$ be a Hermitian form on $Y_1 \times \Xi$, i.e.,

$$F(y, \xi) = (F_1 y, y)_{-1,1} + 2 \operatorname{Re}(F_2 y, \xi)_\Xi + (F_3 \xi, \xi)_\Xi,$$

where

$$F_1 = F_1^* \in \mathcal{L}(Y_1, Y_{-1}), \; F_2 \in \mathcal{L}(Y_0, \Xi),$$
$$F_3 = F_3^* \in \mathcal{L}(\Xi, \Xi).$$

Define the *frequency-domain condition*

$$\alpha := \sup_{\omega, y, \xi} (\|y\|_1^2 + \|\xi\|_\Xi^2)^{-1} F(y, \xi),$$

where the supremum is taken over all triples $(\omega, y, \xi) \in \mathbb{R}_+ \times Y_1 \times \Xi$ such that $i\omega y = Ay + B\xi$.

a) Let, in addition to the above assumption, A be the generator of a C_0-*group* on Y_0 and the pair $(A, -B)$ be L^2-controllable. Then the condition $\alpha \leq 0$, where α is from (**F3**), is sufficient for the application of a theorem by [Likhtarnikov and Yakubovich, 1976]. Note that the existence of C_0-groups is given for conservative wave equations, plate problems, and other important PDE classes [Flandoli *et al.*, 1988] .

3.4   Absolute observation - stability of evolutionary inequalities

We continue the investigation of energy like properties for the observation operators from the inequality problem (3.6), (3.7) with $f \equiv 0$.

The next definition generalizes the concepts which are introduced in [Likhtarnikov and Yakubovich, 1976] for output operators of evolution equations, namely in extending them to the observation operators of a class of evolutionary variational inequalities. In the following we denote for a function $z(\cdot) \in L^2(\mathbb{R}_+; Z)$ their norm by

$$\|z(\cdot)\|_{2,Z} := \left(\int_0^\infty \|z(t)\|_Z^2 \, dt \right)^{1/2} .$$

Definition 3.3.

a) The inequality (3.6), (3.7) is said to be **absolutely dichotomic** (i.e., in the classes $\mathcal{N}(F, G), \mathcal{M}(d)$) **with respect to the observation** z from (3.8) if for any solution $\{y(\cdot), \xi(\cdot)\}$ of (3.6), (3.7) with $y(0) = y_0$, $\xi(0) = \xi_0 \in \mathcal{E}(y_0)$ the following is true: Either $y(\cdot)$ is unbounded on $[0, \infty)$ in the Y_0-norm or $y(\cdot)$ is bounded in Y_0 in this norm and there exist constants c_1 and c_2 (which depend only on $A, B, \mathcal{N}(F, G)$ and $\mathcal{M}(d)$) such that

$$\|Dy(\cdot) + E\xi(\cdot)\|_{2,Z}^2 \leq c_1(\|y_0\|_0^2 + c_2) . \tag{3.9}$$

b) The inequality (3.6), (3.7) is said to be **absolutely stable with respect to the observation** z from (3.8) if (3.9) holds for any solution $\{y(\cdot), \xi(\cdot)\}$ of (3.6), (3.7).

Definition 3.4. The inequality (3.6)−(3.8) with $f \equiv 0$ is said to be **minimally stable** if the resulting equation for $\psi \equiv 0$ is minimally stable, i.e., there exists a bounded linear operator $K : Y_1 \to \Xi$ such that the operator

$A + BK$ is stable, i.e. for some $\varepsilon > 0$

$$\sigma(A + BK) \subset \{s \in \mathbb{C} : \operatorname{Re} s \leq -\varepsilon < 0\}$$

with $\qquad F(y, Ky) \geq 0 \,, \qquad \forall\, y \in Y_1 \,, \qquad\qquad (3.10)$

and $\quad \displaystyle\int_s^t G(y(\tau), \, Ky(\tau)) d\tau \geq 0 \,,$

$$\forall\, s, t : 0 \leq s < t \,, \quad \forall\, y \in L^2_{\mathrm{loc}}(\mathbb{R}_+; Y_1) \,. \qquad (3.11)$$

With the superscript ^{c} we denote the complexification of spaces and operators and the extension of quadratic forms to Hermitian forms.

Theorem 3.1. *Consider the evolution problem* (3.6) – (3.8) *with* $\varphi \in \mathcal{N}(F, G)$ *and* $\psi \in \mathcal{M}(d)$. *Suppose that for the operators* A^c, B^c *the assumptions* (**F1**) *and* (**F2**) *are satisfied. Suppose also that there exist an* $\alpha > 0$ *such that with the transfer operator*

$$\chi^{(z)}(s) = D^c(sI^c - A^c)^{-1}B^c + E^c \qquad (s \notin \sigma(A^c)) \qquad (3.12)$$

the frequency-domain condition

$$F^c\left((i\omega I^c - A^c)^{-1}B^c\xi, \xi\right)$$

$$+ \, G^c\left((i\omega I^c - A^c)^{-1}B^c\xi, \xi\right) \leq -\alpha \|\chi^{(z)}(i\omega)\xi\|^2_{Z^c}$$

$$\forall\, \omega \in \mathbb{R} : i\omega \notin \sigma(A^c) \,, \quad \forall\, \xi \in \Xi^c$$

is satisfied and the functional

$$J(y(\cdot), \xi(\cdot)) := \int_0^\infty \left[F^c(y(\tau), \xi(\tau)) + G^c(y(\tau), \xi(\tau))\right.$$

$$\left. + \, \alpha \|D^c y(\tau) + E^c \xi(\tau)\|^2_{Z^c}\right] d\tau$$

is bounded from above on any set

$$\mathfrak{M}_{y_0} := \{y(\cdot), \xi(\cdot) : \dot{y} = Ay + B\xi \text{ on } \mathbb{R}_+,$$

$$y(0) = \, y_0 \,, \, y(\cdot) \in \mathcal{L}_\infty \,, \, \xi(\cdot) \in L^2(0, \infty; \Xi)\} \,.$$

Suppose further that the inequality (3.6–(3.8) *with* $f \equiv 0$ *is minimally stable, i.e.,* (3.10) *and* (3.11) *are satisfied with some operator* $K \in \mathcal{L}(Y_1, \Xi)$ *and that the pair* $(A + BK, D + EK)$ *is observable in the sense of Kalman* [Brusin, 1976], *i.e., for any solution* $y(\cdot)$ *of*

$$\dot{y} = (A + BK)y \,, \quad y(0) = y_0 \,,$$

with $z(t) = (D + EK)y(t) = 0$ *for a.a.* $t \geq 0$ *it follows that* $y(0) = y_0 = 0$.

Then inequality (3.6), (3.7) *is absolutely stable with respect to the observation* z *from* (3.8).

Proof. Under the assumptions of the given theorem there exist by [Likhtarnikov and Yakubovich, 1976] a (real) operator $P = P^* \in \mathcal{L}(Y_{-1}, Y_0) \cap \mathcal{L}(Y_0, Y_1)$ and a number $\delta > 0$ such that the dissipation inequality is satisfied. Setting in this inequality $\xi = Ky$ from (3.10) with arbitrary $y \in Y_1$ we get with (3.11) the property

$$((A + BK)y, Py)_{-1,1} \leq -\delta \|Dy + EKy\|_Z^2,$$

$$\forall y \in Y_1. \tag{3.13}$$

Using the fact that $A + BK$ is a stable operator and the pair $(A + BK, D + EK)$ is observable, it follows [Brusin, 1976] from (3.13) that $P = P^* \geq 0$. Suppose now that $\{y(\cdot), \xi(\cdot)\}$ is an arbitrary solution of (3.6), (3.7) with $f \equiv 0$. With the Lyapunov-functional $V(y) = (y, Py)_0 \geq 0$ it follows from the dissipation inequality that for arbitrary $t \geq 0$

$$-V(y_0) - \Phi(Cy_0)$$

$$+ \int_0^t [\psi(y(\tau)) - \psi(-Py(\tau) + y(\tau))]d\tau$$

$$+ \delta \int_0^t \|Dy(\tau) + E\xi(\tau)\|_Z^2 \, d\tau \leq 0. \tag{3.14}$$

Since by assumption $\int_0^t [\psi(y(\tau)) - \psi(-Py(\tau) + y(\tau))]d\tau \geq -c_2 > -\infty$ we get from (3.14) for arbitrary $t \geq 0$ the inequalities

$$\delta \int_0^t \|Dy(\tau) + E\xi(\tau)\|_Z^2 \, d\tau$$

$$\leq V(y_0) + \Phi(Cy_0) + c_2$$

$$\leq V(y_0) + c\|y_0\|_0^2 + c_2. \tag{3.15}$$

The properties (3.15) imply now the estimate (3.9).   □

3.5 Application of observation stability to the beam equation

Consider the equation of a beam of length l, with damping and Hookean material, given as

$$\rho\,\mathbf{A}\frac{\partial^2 u}{\partial t^2} + \gamma\frac{\partial u}{\partial t} - \frac{\partial}{\partial x}\left(\frac{\mathbf{EA}}{3}\tilde{g}\left(\frac{\partial u}{\partial x}\right)\right) = 0\,, \tag{3.16}$$

$$u(0,t) = u(l,t) = 0 \quad \text{for} \quad t > 0\,, \tag{3.17}$$

$$u(x,0) = u_0(x)\,, \ u_t(x,0) = u_1(x) \tag{3.18}$$

$$\text{for} \quad x \in (0,l)\,.$$

Here u is the deformation in the x direction. Assume that the cross section area $\mathbf{A}$, the viscose damping γ, the mass density ρ and the generalized modulus of elasticity $\mathbf{E}$ are constant. The nonlinear stress-strain law $\tilde{g}$, is given by

$$\tilde{g}(w) = 1 + w - (1 + w)^{-2}\,, \quad w \in (-1,1)\,. \tag{3.19}$$

Let us break the stress-strain law into the sum of a linear term and a nonlinear term as $\tilde{g}(w) = g(w) + w$. Then the above model (3.16) can be rewritten as

$$\rho\mathbf{A}\frac{\partial^2 u}{\partial t^2} - \frac{\partial}{\partial x}\left(\frac{\mathbf{EA}}{3}\frac{\partial u}{\partial x}\right)$$
$$+ \gamma\frac{\partial u}{\partial t} - \frac{\partial}{\partial x}\left(\frac{\mathbf{EA}}{3}g\left(\frac{\partial u}{\partial x}\right)\right) = 0\,. \tag{3.20}$$

Assume the Gelfand triple $\mathcal{V}_1 \subset \mathcal{V}_0 \subset \mathcal{V}_{-1}$ with

$$\mathcal{V}_0 := L^2(0,l)\,, \quad \mathcal{V}_1 := H_0^1(0,l)$$
$$\text{and} \quad \mathcal{V}_{-1} := H^{-1}(0,l)\,. \tag{3.21}$$

Then equation (3.16) − (3.18) can be rewritten in $\mathcal{V}_{-1}$ as

$$\rho\mathbf{A}u_{tt} + \mathcal{A}_1 u + \mathcal{A}_2 u_t + \mathcal{C}^* g(\mathcal{C}u) = 0\,, \tag{3.22}$$

$$u(0) = u_0\,, \quad u_t(0) = u_1\,, \tag{3.23}$$

with $\mathcal{A}_1 \in \mathcal{L}(\mathcal{V}_1, \mathcal{V}_{-1})$, $\mathcal{A}_2 \in \mathcal{L}(\mathcal{V}_1, \mathcal{V}_{-1})$ (strong damping), $\mathcal{C} \in \mathcal{L}(\mathcal{V}_1, \mathcal{V}_0)$ and $g : \mathcal{V}_0 \to \mathcal{V}_0$. The operators $\mathcal{A}_1$ and $\mathcal{A}_2$ are associated with their bilinear forms $a_i : \mathcal{V}_1 \times \mathcal{V}_1 \to \mathbb{R}$ ($i = 1, 2$) through $(\mathcal{A}_i v, w)_{\mathcal{V}_{-1}, \mathcal{V}_1} = a_i(v, w)$, $\forall\, v, w \in \mathcal{V}_0$.

In order to get a variational interpretation of (3.22), (3.23) we make the following assumptions [Banks *et al.* 1997; Banks and Ito, 1988] :

(A1)

a) The form a_1 is symmetric on $\mathcal{V}_0 \times \mathcal{V}$;

b) a_1 is $\mathcal{V}_1$ continuous, i.e., for some $c_1 > 0$ holds $|a_1(v,w)| \leq c_1\|v\|_{\mathcal{V}_1}\|w\|_{\mathcal{V}_1}$, $\forall v, w \in \mathcal{V}_1$;

c) a_1 is strictly $\mathcal{V}_1$-elliptic, i.e., for some $k_1 > 0$ holds $a_1(v,v) \geq k_1\|v\|_{\mathcal{V}_1}^2$, $\forall v \in \mathcal{V}_1$.

(A2)

a) The form a_2 is $\mathcal{V}_1$ continuous, i.e., for some $c_2 > 0$ holds $|a_2(v,w)| \leq c_2\|v\|_{\mathcal{V}_1}\|w\|_{\mathcal{V}_1}$, $\forall v, w \in \mathcal{V}_1$.

b) The form a_2 is $\mathcal{V}_1$ coercive and symmetric, i.e., there are $k_2 > 0$ and $\lambda_0 \geq 0$ s.t.

$$a_2(v,v) + \lambda_0\|v\|_{\mathcal{V}_0}^2 \geq k_2\|v\|_{\mathcal{V}_1}^2 \qquad \text{and}$$
$$a_2(v,w) = a_2(w,v), \quad \forall v, w \in \mathcal{V}_1.$$

(A3)

a) The operator $\mathcal{C} \in \mathcal{L}(\mathcal{V}_1, \mathcal{V}_0)$ satisfies with some $k \geq 0$ the inequality

$$\|\mathcal{C}v\|_{\mathcal{V}_0} \leq \sqrt{k}\|v\|_{\mathcal{V}_1}, \quad \forall v \in \mathcal{V}_1.$$

$g : \mathcal{V}_0 \to \mathcal{V}_0$ is continuous and $\|g(v)\|_{\mathcal{V}_0} \leq c_1\|v\|_{\mathcal{V}_0} + c_2$ for $v \in \mathcal{V}_0$, where c_1 and c_2 are nonnegative constants.

b) g is of gradient type, i.e., there exists a coninuous Frechét-differentiable functional $G : \mathcal{V}_0 \to \mathbb{R}$, whose Frechét derivative $G'(v) \in \mathcal{L}(\mathcal{V}_0, \mathbb{R})$ at any $v \in \mathcal{V}_0$ can be represented in the form

$$G'(v)w = (g(v), w)_{\mathcal{V}_0}, \quad \forall w \in \mathcal{V}_0.$$

c) $g(0) = 0$ and for some positive $\varepsilon < 1$ we have for all $v, w \in \mathcal{V}_0$

$$(g(v) - g(w), v - w)_{\mathcal{V}_0}$$
$$\geq -\varepsilon k_1 k^{-1}\|v - w\|_{\mathcal{V}_0}^2. \tag{3.24}$$

We say that $u \in \mathcal{L}_T$ is a *weak solution* of (3.22), (3.23) if

$$(u_{tt}, \eta)_{\mathcal{V}_{-1}, \mathcal{V}_1} + a_1(u, \eta) + a_2(u_t, \eta) \tag{3.25}$$
$$+ (g(\mathcal{C}u), \mathcal{C}u)_0 = 0 \qquad \forall \eta \in \mathcal{L}_T, \text{ a.a. } t \in [0, T].$$

Let us formulate our problem (3.25) in first order form on the energetic space $Y_0 := \mathcal{V}_1 \times \mathcal{V}_0$ in the coordinates $y = (y_1, y_2) = (u, u_t)$. Define for this $Y_1 := \mathcal{V}_1 \times \mathcal{V}_1$ and $a : Y_1 \times Y_1 \to \mathbb{R}$ by

$$a((v_1, v_2), (w_1, w_2))$$
$$= (v_2, w_1)_{\mathcal{V}_1} - a_1(v_1, w_2) - a_2(v_2, w_2),$$
$$\forall (v_1, v_2), (w_1, w_2) \in Y_1 \times Y_1. \tag{3.26}$$

The norms in the product spaces Y_0 and Y_1 are given in the standard way by

$$\|(y_1, y_2)\|_0^2 := \|y_1\|_{\mathcal{V}_1}^2 + \|y_2\|_{\mathcal{V}_0}^2, \quad (y_1, y_2) \in Y_0,$$

and

$$\|(y_1, y_2)\|_1^2 := \|y_1\|_{\mathcal{V}_1}^2 + \|y_2\|_{\mathcal{V}_1}^2, \quad (y_1, y_2) \in Y_1.$$

Then (3.25) can be rewritten as

$$(\dot{y}, \eta)_{-1,1} - a(y, \eta) = (B\varphi(Cy), \eta)_{-1,1}, \tag{3.27}$$
$$y(0) = (u_0, u_1), \quad \forall \eta \in Y_1,$$

where

$$B\varphi(Cy) := \begin{pmatrix} 0 \\ -\mathcal{C}^* g\,(\mathcal{C}y_1) \end{pmatrix}. \tag{3.28}$$

We can also write (3.27), (3.28) formally in the operator form

$$\dot{y} = Ay + B\varphi\,(Cy), \ y(0) = y_0, \tag{3.29}$$

where A is defined by

$$a(v, w) = (Av, w)_{-1,1}, \quad \forall v, w \in Y_1,$$

i.e., $A = \begin{bmatrix} 0 & I \\ -\mathcal{A}_1 & -\mathcal{A}_2 \end{bmatrix}$.

It is shown in [Banks *et al.* 1997; Banks and Ito, 1988] that the embedding $Y_1 \subset Y_0$ is completely continuous and the operator A generates an analytic semigroup on Y_1, Y_0 and $Y_{-1} = \mathcal{V}_1 \times \mathcal{V}_{-1}$. Furthermore, its semigroup is exponentially stable on Y_1, Y_0 and Y_{-1}. From this it follows that the pair (A, B) is exponentially stabilizable. Let us consider with parameters $\varepsilon > 0$ and $\alpha \in \mathbb{R}$ a more simplified form of (3.16) − (3.18) written as

$$\frac{\partial^2 u}{\partial t^2} + 2\varepsilon \frac{\partial u}{\partial t} - \alpha \frac{\partial^2 u}{\partial x^2} \tag{3.30}$$
$$= -\alpha \left(\frac{\partial}{\partial x} \left(-g\left(\frac{\partial u}{\partial x} \right) \right) \right) =: \alpha \frac{\partial}{\partial x} \xi$$

together with the boundary and initial conditions (3.17), (3.18), where we have $\xi = -g = \varphi$ introduced as new nonlinearity. According to (3.24) in (**A3**)a) we can assume that $\varphi \in \mathcal{N}(F)$ with the quadratic form $F(w, \xi) = \mu w^2 - \xi w$ on $\mathbb{R} \times \mathbb{R}$, where $\mu > 0$ is a certain parameter. Note that it is possible to include a second quadratic form G if we use the information from (**A3**)b).

Suppose that $\lambda_k > 0$ and e_k ($k = 1, 2, \ldots$) are the eigenvalues resp. eigenfunctions of the operator $-\Delta$ with zero boundary conditions. We write formally the Fourier series of the solution $u(x,t)$ and the perturbation $\xi(x,t)$ to the (linear) equation (3.30) as

$$u(x,t) = \sum_{k=1}^{\infty} u^k(t)e_k \quad \text{and} \quad \xi(x,t) = \sum_{k=1}^{\infty} \xi^k(t)e_k. \tag{3.31}$$

If we introduce the Fourier transforms $\tilde{u}$ and $\tilde{\xi}$ of (3.31) with respect to the time variable we get from (3.30) for $k = 1, 2, \ldots$ the equations

$$-\omega^2 \tilde{u}^k(i\omega) + 2i\omega\varepsilon\tilde{u}^k(i\omega) + \lambda_k \tilde{u}^k(i\omega)$$
$$= -\alpha\sqrt{\lambda_k}\,\tilde{\xi}^k(i\omega). \tag{3.32}$$

It follows from (3.32) that for $k = 1, 2, \ldots$

$$\tilde{u}^k = \chi(i\omega, \lambda_k)\tilde{\xi}^k, \tag{3.33}$$

where

$$\chi(i\omega, \lambda_k) = (-\omega^2 + 2i\omega\varepsilon + \alpha\lambda_k)^{-1}(\alpha\lambda_k),$$
$$\forall\, \omega \in \mathbb{R} : -\omega^2 + 2i\omega\varepsilon + \alpha\lambda_k \neq 0. \tag{3.34}$$

In order to check the sufficient conditions for Theorem 3.1 we consider the functional

$$J(w, \xi) := \operatorname{Re} \int_0^{\infty} \int_0^l (\mu|w|^2 - w\xi^*)\,dx\,dt. \tag{3.35}$$

Using the Parseval equality for (3.35) with

$$|\tilde{w}|^2 = \sum_{k=1}^{\infty} \lambda_k |\tilde{u}^k|^2 = \sum_{k=1}^{\infty} \lambda_k |\tilde{u}^k|^2$$
$$= \sum_{k=1}^{\infty} \lambda_k |\chi(i\omega, \lambda_k)|^2 |\tilde{\xi}^k|^2$$

and

$$\tilde{w}\,\tilde{\xi}^* = \sum_{k=1}^{\infty} \sqrt{\lambda_k}\,\tilde{u}^k\,(\tilde{\xi}^k)^* = \sum_{k=1}^{\infty} \sqrt{\lambda_k}\,\chi\,(i\omega, \lambda_k)|\tilde{\xi}^k|^2,$$

we conclude [Arov and Yakubovich, 1982] that the functional (3.35) is bounded from above if and only if the functional

$$\mathrm{Re} \int_{-\infty}^{+\infty} \int_0^l [\mu \left(\sum_{k=1}^{\infty} \lambda_k |\chi(i\omega, \lambda_k)|^2 |\tilde{\xi}^k|^2 \right.$$

$$- \sum_{k=1}^{\infty} \sqrt{\lambda_k}\,\chi\,(i\omega, \lambda_k)|\tilde{\xi}^k|^2 \,]\,dx d\omega \qquad (3.36)$$

is bounded on the subspace of Fourier-transforms defined by (3.33), (3.34) or, using again a result of [Arov and Yakubovich, 1982], that the frequency-domain condition

$$\mu\lambda_k|\chi\,(i\omega, \lambda_k)|^2 - \sqrt{\lambda_k}\,\mathrm{Re}\,\chi(i\omega, \lambda_k) < 0\,, \qquad (3.37)$$

$$\forall\,\omega \in \mathbb{R} : -\omega^2 + 2i\omega\varepsilon + \alpha\lambda_k \neq 0\,,\ k = 1, 2, \ldots,$$

is satisfied, where $\chi\,(i\omega, \lambda_k) = (-\omega^2 + 2i\omega\varepsilon + \alpha\lambda_k)^{-1}(-\alpha\sqrt{\lambda_k})$. Clearly, (3.37) describes a certain domain Q in the space of parameters $\mu >$ $0, \varepsilon > 0, \alpha \in \mathbb{R}$. Theorem 3.1 shows that (3.29), associated with (3.30),(3.17),(3.18) is absolutely stable with respect to the observation $z = (y_1, y_2)$, if the parameter from Q also guarantee the minimal stability of (3.29).

References

Arov, D. Z. and Yakubovich, V. A. (1982). Conditions for semiboundedness of quadratic functionals on Hardy spaces. *Vestn. Leningr. Univers. Ser. Mat., Mekh., Astr.*, **1**, pp. 11–18 (in Russian).

Banks, H. T., Gilliam, D. S. and Shubov, V. (1997). Global solvability for damped abstract nonlinear hyperbolic systems. *Differential and Integral Equations*, **10**, pp. 309–332.

Banks, H. T. and Ito, K. (1988). A unified framework for approximation in inverse problems for distributed parameter systems. *Control-Theory and Advanced Technology*, **4**, pp. 73–90.

Brézis, H. (1972). Problemes unilateraux. *J. Math. Pures Appl.*, **51**, pp. 1–168.

Brusin, V. A. (1976). The Luré equations in Hilbert space and its solvability. *Prikl. Math. Mekh.*, **40** (5), pp. 947–955 (in Russian).

Duvant, G. and Lions, J.-L. (1976). *Inequalities in Mechanics and Physics.* Springer-Verlag, Berlin.

Flandoli, F., Lasiecka J. and Triggiani, R. (1988). Algebraic Riccati equations with nonsmoothing observation arising in hyperbolic and Euler-Bernoulli boundary control problems. *Annali di Matematica Pura Applicata,* **153**, pp. 307–382.

Yakubovich, V. A., Leonov, G. A. and Gelig, A. Kh. (2004). *Stability of Stationary Sets in Control Systems with Discontinuous Nonlinearities.* World Scientific, Singapore.

Han, W. and Sofonea, M. (2000). Evolutionary variational inequalities arising in viscoelastic contact problems. *SIAM J. Numer. Anal.,* **38** (2), pp. 556–579.

Kuttler, K. L. and Shillor, M. (1999). Set-valued pseudomonotone maps and degenerate evolution inclusions. *Comm. Contemp. Math.,* **1** (1), pp. 87–123.

Likhtarnikov, A. L. and Yakubovich, V. A. (1976). The frequency theorem for equations of evolutionary type. *Siberian Math. J.,* **17** (5), pp. 790–803.

Pankov, A. A. (1986). *Bounded and almost periodic solutions of nonlinear differential operator equations.* Naukova dumka, Kiev (in Russian).

Reitmann, V. (2005). Convergence in evolutionary variational inequalities with hysteresis nonlinearities. *In: Proceedings of Equadiff 11, 2005, Bratislava,* pp. 395–404.

A Discrete-time Hybrid Lurie Type System with Strange Hyperbolic Nonstationary Attractor

V.N. Belykh, B. Ukrainsky

Volga State Academy of Water Transport,
Nizhniy Novgorod,
Russia

Abstract

In this paper we present sufficient conditions for existence of strange hyperbolic nonstationary attractor of hybrid continuous piecewise smooth discrete-time dynamical system.

4.1 Introduction

In the field of dynamical chaos hyperbolic strange attractors play a central role as one of the basic units linking dynamical and ergodic theories. Hyperbolic strange attractors generate random (in terms of mixing property) stationary (in terms of Sinai-Bowen-Ruelle measure (SBR-measure)) processes [Anosov and Sinai, 1967; Bowen, 1977; Katok and Hasselblatt, 1995; Afraimovich *et al.*, 1995]. In the case of ODEs there are several examples showing the possible existence of a hyperbolic attractor [Belykh *et al.*, 2005; Kuznetsov, 2005; Kuznetsov and Pikovsky, 2007]. Unfortunately all these examples do not lend itself so far to mathematical verification. Contrary, in the case of maps (discrete-time dynamical systems) there are several well defined examples (Smale-Williams attractor, Lozi map, Belykh map, etc) [Katok and Hasselblatt, 1995; Lozi, 1978; Belykh, 1995; Belykh *et al.*, 2002] for which the hyperbolicity, the existence of invariant measure and mixing property are proved [Pesin, 1992; Sataev, 1999; Schmeling and Troubetzkoy, 1998]. In particular, the first example of hyperbolic attractor in Lurie discrete-time system with continuous nonlinearity, having a bounded away from zero discontinuous derivative, was presented in [Belykh *et al.*, 2002]. This system serves as a model of electro-mechanical control systems. In the map representation it takes the form

$$\bar{u} = Au + p\varphi(x), \quad x = c^T u, \tag{4.1}$$

where $u = u(i)$, $\bar{u} = u(i+1)$, $u \in \mathbb{R}^m$; A is constant $m \times m$-matrix; p, c are constant $m \times 1$-vectors; "T" is transpose and $\varphi : \mathbb{R}^1 \to \mathbb{R}^1$ is a continuous piecewise smooth function, $|\varphi'(x)| > K$, $x \in \mathbb{R}^1$.

For the parameter domain [Belykh *et al.*, 2002] where the map (4.1) is hyperbolic from papers [Sataev, 1999; Schmeling, 1998] it follows that the hyperbolic attractor of (4.1) is stationary in the sense of SBR-measure.

In the present paper we consider the hybrid system of the form

$$\begin{cases} u(i+1) = Au(i) + p\varphi(x(i), z(i)) \\ z(i+1) = \psi(x(i), z(i)), \quad x = c^T u, \end{cases} \tag{4.2}$$

where the integer $z \in \mathbb{Z}$, the function $\psi : \mathbb{R}^1 \times \mathbb{Z}_N \to \mathbb{Z}_N$ is bounded: $[1, N] = \mathbb{Z}_N$.

Our main purpose is to obtain sufficient conditions such that the hybrid map (4.2) has a strange hyperbolic nonstationary attractor.

Note, that in the case of a periodic nonautonomous system (4.2) when $\psi = z(\mod N)$, the system (4.2) is a composition of N maps (4.1) with $\varphi(x, i)$, $i = 0, 1, ..., N-1$ standing for $\varphi(x)$, and the hyperbolicity of each such map from N sequential maps does not imply that the map (4.2) is also hyperbolic. We consider an arbitrary (even random) sequence of $z(i)$ generated by the second equation in (4.2). The proof of hyperbolicity is based on a comparison principle for multidimensional maps, and the construction of cones that are invariant with respect to a linearization of the map (4.2), and are independent of the phase coordinates.

4.2   Reduction to a normal form

We introduce the normal form of the Lurie system as the following map F

$$\begin{cases} \begin{pmatrix} \bar{x} \\ \bar{y} \end{pmatrix} = \begin{pmatrix} 1 & 1 \\ 0 & B \end{pmatrix} \begin{pmatrix} x \\ y \end{pmatrix} - \begin{pmatrix} a \\ b \end{pmatrix} g(x, z) \\ \bar{z} = \psi(x, z), \end{cases} \tag{4.3}$$

where the overbar denotes the forward shift in time, $g(x, z) \equiv kx + \varphi(x, z)$; B is $n \times n$-matrix $(n = m - 1)$, $y = column(y_1, y_2, \ldots, y_n)$, $\mathbf{1} = (1, 1, \ldots, 1)$, $b = column(b_1, b_2, \ldots, b_n)$ is n-vector of parameters, k and a are scalar parameters. Denoting $v - (x, y^T)^T$, and introducing the transformation $v = Su$, where S is a nonsingular $m \times m$-matrix, we obtain that the system (4.2) takes the form (4.3) as long as the following system of equations has a solution:

$$
\begin{cases}
SAS^{-1} = \begin{pmatrix} 1 - ka & 1 \\ -kb & B \end{pmatrix} \\
c^T S^{-1} = e^T \\
Sp = \begin{pmatrix} -a \\ -b \end{pmatrix},
\end{cases}
\tag{4.4}
$$

where $e^T = (1, 0, \ldots, 0)$. Note, that the following system of equations

$$
\begin{cases}
kp + (E - SAS^{-1})e = 0 \\
c^T S^{-1} = e^T \\
e^T SAS^{-1} = \begin{pmatrix} 1 - ka & 1 \end{pmatrix}
\end{cases}
\tag{4.5}
$$

takes a form of necessary conditions for resolving the system (4.4). We assume that the system (4.5) can be resolved with respect to the parameter k and matrix S (an example is shown in [Belykh *et al.*, 2002]). Hence, the system (4.1) is reduced to the map (4.3) which we consider below.

4.3   Existence of invariant domain

Consider an arbitrary map $\Phi : \mathbb{R}^{m+n} \to \mathbb{R}^{m+n}$ of the form $(x, y) \to (P(x, y), Q(x, y))$, and reduced map $\Phi_0 : (x, y) \to (P(x, y), y)$, where $x \in \mathbb{R}^m$, $y \in \mathbb{R}^n$ and $y = const$ for Φ_0. Our problem is to derive the conditions for the map Φ as well as for the boundaries of a domain D such that 1) $\Phi D \subset D$; 2) $D = D_x \times D_y$ (direct product).

Comparison principle. Assume that there exist some compacts D_x and D_y such that:

(1) $\Phi_0 D_x \subset D_x$ for any $y = const$ from compact D_y
(2) $Q(x, y) \in D_y$ for any $x \in D_x$ and $y \in D_y$.

Then D is invariant under the map Φ.

From this principle it follows that the map has an attractor $A = \Phi A, A \subset D$.

Remark. The variables separation in this obvious principle is immediately directed to the finding of the compacts D_x and D_y for certain maps. The map, which we consider in the paper is the case.

As the main example we consider the following class of nonlinear functions. For a natural $n > 1$, from the interval $[c, d]$ consider two sets of real numbers $S_a = (a_0 = c < a_1 < a_2 < \ldots < a_n = d)$ and $S_b = (b_0 = 0, b_1, \ldots, b_n = 0) | b_{i-1} b_i < 0$, $i = \overline{2, n-1}$, and consider a set of functions $S_\eta = (\eta_1(\xi), \eta_2(\xi), \ldots, \eta_n(\xi))$, where $\eta_1(\xi)$ is given in the interval $(-\infty, a_1]$, $\eta_n(\xi)$ — in the interval $[0, \infty)$, and for $i = \overline{2, n-1}$ the functions $\eta_i(\xi)$ are given in the intervals $[0, a_i - a_{i-1}]$. Assume that each function from S_η is continuous, smooth; $\eta_i(0) = 0$, $\eta_i'(\xi) > k > 0$. Introduce the following function $\eta(x, n) = b_{i-1} + \dfrac{b_i - b_{i-1}}{\eta_i(a_i - a_{i-1})} \eta_i(x - a_{i-1})$, where index $i = 1$ for $x \in (-\infty, a_1]$, index $i = n$ for $x \in [a_n, \infty)$ and for $x \in (a_{i'-1}, a_{i'}]$ index $i = i'$, $i = \overline{2, n-1}$. This function has singularities at critical points a_i and $f(a_i) = b_i$ for $i = \overline{1, n-1}$. An example of such function for $n = 5$, $a_i = c + (i-1)\dfrac{d-c}{n}$, $b_i = (-1)^i b$, $\eta_i(x) = x$ for all i is depicted in Fig. 4.1.

Now the map (4.3) is defined with the function $g(x, z) = \eta(x, z)$ and an arbitrary function $\psi(x, z)$. We consider a set of functions $\Re(h) : f(x, z) = x - ag(x, z)$, such that:

(1) For even z $M = \max\limits_{x \in [c, d]} f(x, z) < d$; $m = \min\limits_{x \in [c, d]} f(x, z) > c$. For odd z
$f(M, z) > c$, $f(m, z) > c$
(2) $\max\limits_{x \in [c, d]} f'(x, z) > h$.

Figure 4.2 illustrates functions $f(x, z) \in \Re(h)$ for even and odd z.

First we consider reduced system (4.3), that is a one-parameter z family of maps F_1:

$$\begin{cases} \overline{x} = x + 1y - ag(x, z) \\ \overline{y} = By - bg(x, z), \end{cases} \qquad (4.6)$$

where $z \in \mathbb{Z}_N$ is a constant parameter. Under a nonsingular linear transformation the map (4.6) can be reduced to the form

$$\begin{cases} \overline{x} = x + 1y - ag(x, z) \\ \overline{y}_i = \lambda_i(y_i - b_i g(x, z)), \end{cases} \qquad (4.7)$$

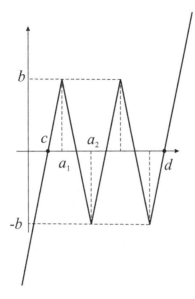

Fig. 4.1 Example of $\eta(x, 4)$ with 4 critical points.

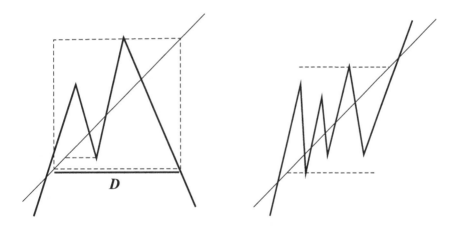

Fig. 4.2 An example $f(x, z) \in \Re(h)$ for even and odd z.

where λ_i denotes either a real eigenvalue of matrix B or $\alpha \pm \beta$ for complex eigenvalues $\alpha \pm \beta i$ of matrix B. Note that in the case of multiple eigenvalues of matrix B some of the such values of λ_i must be increased by some quantity ε from a Jordan form of the matrix B.

Applying the comparison principle for the map F_1 as the auxiliary map Φ_0 we consider the map F_0 in the form $(x, y) \to (x - ag(x, z) + Y, y)$, where $Y = 1y$ a parameter. This map is a two parameter $y = const, z = const$ family of one dimensional maps $f_1 : \overline{x} = x - ag(x, z) + Y$. As for each $f_1 \in \Re(h)$ the first condition of the comparison principle is fulfilled, so the interval $[c, d] = D_x$. This condition is illustrated in Fig. 4.2 for our main example. It is easy to verify that there exists an interval $[y^-, y^+] = D_y$, satisfying the condition 2 of the comparison principle. In fact this condition becomes valid for small enough eigenvalues of the matrix B.

These conditions provide a simple technical rule for the system (4.3): from $x \in [c, d]$ and $Y \in [y^-, y^+]$ it follows that $\overline{x} \in [c, d]$. Under the condition $F_1 D \subset D$ the next theorem holds.

Theorem 4.1. *There exists a number λ_0, such that for any $\lambda^+ < \lambda_0$ $(\lambda^+ \overset{\Delta}{=} \max_i\{|\lambda_i|\})$ and any $z \in \mathbb{Z}_N$ the map F_1 has an invariant domain $D = \{(x, y) : c < x < d, y_i^- < y_i < y_i^+, i = \overline{1, n}\}$ and, therefore this map has an attractor $A \subset D$.*

4.4 Conditions of hyperbolicity

Assume, that $g(x, z)$ is a continuous piecewise smooth function, i.e. smooth in each interval of monotonicity. Denote $h \overset{\Delta}{=} \inf_{x \in A, z \in \mathbb{Z}_N} |g'_x(x, z)|$.

Definition 4.1. We call a cone in R^{n+1} with one-dimensional axes being a set of vectors of the form $K_1 = \{(\xi, \eta) \in R^{n+1} : \xi \in R^1, \eta \overset{\Delta}{=} column(\eta_1, \eta_2, ... \eta_n) \in R^n, \frac{\eta_i}{\xi} = \alpha_i, \alpha_i \in (\alpha_i^-, \alpha_i^+), i = \overline{1, n}\}$, and we call a cone with n-dimensional axial space $K_n = \{(\xi, \eta) \in R^{n+1} : \xi + \sum_i \beta_i \eta_i = 0, \beta_i \in (\beta_i^-, \beta_i^+), i = \overline{1, n}\}$.

The cone K_1 is a set of vectors being parallel to those vectors, which have one unit coordinate and all the others are bounded, the cone K_n is a set of vectors from n-dimensional plane and its vectors $column(1, \beta_1, ..., \beta_n)$ are similar to those as for K_1 (see Fig. 4.3).

For each $z = const$ consider the linearization of the map F_1 in a point (x, y) of the phase space resulting in the linear map T of the form

$$\begin{cases} \overline{\xi} = (1 - s(x, z))\xi + \sum \eta_i \\ \overline{\eta}_i = -t_i(x, z)\xi + \lambda_i \eta_i \quad i = \overline{1, n}, \end{cases} \tag{4.8}$$

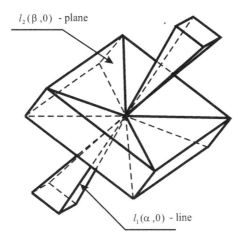

Fig. 4.3 Cones K_1 and K_2.

where $s(x,z) \overset{\Delta}{=} ag'_x(x,z),\quad t_i(x,z) \overset{\Delta}{=} \lambda_i b_i g'_x(x,z)$. We consider the cones in the space (ξ, η), which are independent of the points (x,y) in the phase space of F_1.

Introduce two families of linear manifolds

$$l_1(\alpha,\sigma) \overset{\Delta}{=} \left\{ \begin{array}{l} \eta_i - \alpha_i\xi = \sigma_i \\ i = \overline{1,n}, \end{array} : \begin{array}{l} \sigma_i \in R^1 \\ \alpha_i \in (\alpha_i^-, \alpha_i^+) \end{array} \right\},$$

$$l_2(\beta,\rho) \overset{\Delta}{=} \{\xi + \sum_i \beta_i\eta_i = \rho : \rho \in R^1, \beta_i \in (\beta_i^-, \beta_i^+)\}.$$

The images of these manifolds Tl_1 and Tl_2 are also linear manifolds with new values $\overline{\alpha}_i$, $\overline{\beta}_i$, $\overline{\sigma}_i$ and $\overline{\rho}$.

Definition 4.2. An attractor A is called a hyperbolic attractor, if there exist cones K_1^u and K_n^s, such that $\forall (x,y) \in A$ the following conditions hold:

h1. $clos(K_1^u) \cap clos(K_n^s) = \{0\}$;

h2. $clos(TK_1^u) \subset K_1^u,\quad clos(T^{-1}K_n^s) \subset K_n^s$;

h3. There exist a constant $l, 0 < l < 1$ such, that

    a) If $(\xi,\eta) \in K_1^u$ and $\beta_i \in (\beta_i^-, \beta_i^+)$ then $\overline{\beta}_i \in (\beta_i^-, \beta_i^+)$ and $|\overline{\rho}| > l^{-1}|\rho|$,

    b) If $(\xi,\eta) \in K_n^s$ and $\alpha_i \in (\alpha_i^-, \alpha_i^+)$ then $\overline{\alpha}_i \in (\alpha_i^-, \alpha_i^+)$ and $|\overline{\sigma}_i| < l|\sigma_i|$.

We pay special attention at the condition h3, which implies that the image of a line and pre-image of a plane tend to the origin due to decrease of values ρ and $\overline{\sigma}_i$ in a geometrical progression with the factor l. One iterate of $l_1(\alpha,\sigma)$ and $l_2(\beta,\rho)$ in the cones is schematically shown in the Fig. 4.4.

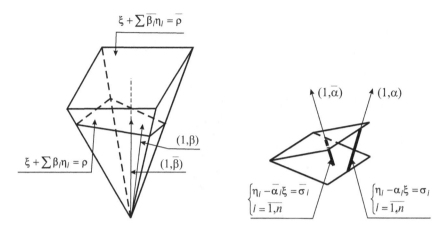

Fig. 4.4   Transformation of linear manifolds.

Under the above conditions on function $g(x, z)$ the following theorem holds.

Theorem 4.2. *There exist numbers $h_0 > 1$ and λ_0, $0 < \lambda_0 < 1$, such that the attractor A of the map F_1 is a hyperbolic attractor for $h > h_0$, $\lambda^+ < \lambda_0$ and any $z \in \mathbb{Z}_N$.*

Proof (in the case of K_1^u) is based on:

(1) Invariance:

   (a) We prove, that the images of $(\alpha_i, \beta_i, \sigma_i, \rho)$ are defined by the next formulas $\beta_i = \dfrac{1 + \overline{\beta}_i \lambda_i}{1 - s - \sum \overline{\beta}_j t_j}$, $\overline{\alpha_i} = \dfrac{-t_i \lambda_i + \lambda_i \alpha_i}{1 - s + \sum \alpha_j}$,
$\overline{\rho} = \rho(1 - s - \sum \overline{\beta}_i t_i)$, $\overline{\sigma}_i = \sigma_i \lambda_i - \overline{\alpha}_i \sum \sigma_j$. We construct a map for the value $\zeta \overset{\Delta}{=} \sum \alpha_i$ generated by the map T. The inequality for the image $\overline{\zeta}$ holds: $f_2(\zeta) < \overline{\zeta} < f_1(\zeta)$ where the functions $f_1(\zeta) = \dfrac{\lambda h + \lambda^- \zeta}{1 + ah + \zeta}$ and $f_2(\zeta) = \dfrac{-\lambda h + \lambda^- \zeta}{1 - ah + \zeta}$ are comparison functions. The graphs of the functions $\overline{\zeta} = f_1(\zeta)$ and $\overline{\zeta} = f_2(\zeta)$ are shown in the Fig.4.5. This figure illustrate the existing invariant interval (ζ^-, ζ^+).

   (b) The existence of an invariant interval implies that there exists a set of intervals for each coordinate α_i. The latter finishes the proof of the existence and invariance of the cone (K_1^u).

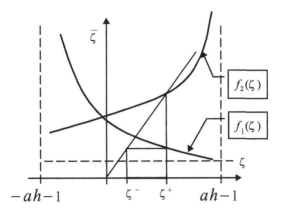

Fig. 4.5   A comparison of one-dimensional maps.

(2) Expansion:

We prove the property of expansion of the variable $\rho \overset{\Delta}{=} \xi + \sum \beta_i \eta_i$. This fact together with boundedness of coordinates of the vector β provides an expansion of any vector $(\xi, \eta) \in K_1^u$.

The proof for invariant cone K_n^s is similar to that for cone K_1^u.

Theorem 4.3. *Let the conditions of Theorem 4.2 hold. Then the map (4.3) has a strange hyperbolic nonstationary attractor.*

Proof. Let two different maps $F_1^{(1)}$ and $F_1^{(2)}$ satisfy the Theorem 4.2. Then the composition $F_1^{(1)} F_1^{(2)}$ satisfies the Theorem 4.2 as well. This statement follows from the property that both $F_1^{(1)}$ and $F_1^{(2)}$ have the same invariant domain and the same invariant cones K_1^u and K_n^s which are the same for any point. At each two neighbor iterate i and $i + 1$ the integer z_i and z_{i+1} takes values within the interval $\mathbb{Z}_N$ and the maps $F \mid_{z=z_i}$ and $F \mid_{z=z_{i+1}}$ are representatives of the family F_1. Hence $F \mid_{z=z_i} \cdot F \mid_{z=z_{i+1}}$ satisfy the Theorem 4.2, the map F has strange hyperbolic nonstationary attractor with respect to integer z. $\square$

Example. For the function from the main example we obtain N different maps (4.3) for the sequence $f(x,1)$, $f(x,2)$,..., $f(x,N)$ from $\Re(h)$. Let the control rule for the integer z be given by the function $\psi(x,y,z) = k \in \mathbb{Z}_N$ with probability p_k, $\sum_{k=1}^{N} p_k = 1$. Due to Theorem 4.3

the map (4.3) has a hyperbolic attractor which randomly changes its structure according to the probability distribution.

Acknowledgments

This work was supported by the NWO-RFBR (grant N 047-017-018) and Russian Foundation for Basic Research (grant N 09-01-00498-a).

References

Afraimovich, V., Chernov, N. and Sataev, E. (1995). Statistical properties of 2-D generalized attractors, *J. of Chaos*, **5**(1), pp. 239-252.

Anosov, D. and Sinay, J. (1967). Some smooth dynamical systems, *UMN*, **22**(5), pp. 107-172.

Belykh, V. (1995). Chaotic and strange attractors of a two-dimensional map, *Matematicheski Sbornik*, **186**(3), pp. 3-18.

Belykh, I., Belykh, V. Mosekilde, E. (2005). Hyperbolic Plykin attractor can exists in neuron models. *Int. J. Bifurc. Chaos*,15, pp. 3567-3578.

Belykh, V., Komrakov, N. and Ukrainsky, B. (2002). Hyperbolic attractors in a family of multidimensional maps with cusp-points. In *Proc. of int. conf. "Progress in nonlinear science" dedicated to the 100-th anniversary of A. Andronov*. Nizhny Novgorod. V. 1, pp. 31-38.

Bowen, R. (1977). Bernoulli maps of the interval. *Israel J. of Math..* **28**, pp.161-168.

Katok, A, Hasselblatt, B. (1995). *Introduction to the Modern Theory of Dynamical Systems*. Cambridge University Press.

Kuznetsov, S. (2005). Example of a physical system with a hyperbolic attractor of the Smale-Williams type. *Phys. Rev. Lett..* 95. p. 14401.

Kuznetsov, S. and Pikovsky, A. (2007). Autonomous coupled oscillators with of a physical system with a hyperbolic strange attractors. *Physica D..* 232. pp. 67-102.

Lozi, R. (1978). Un attracteur de Hénon. *J. Physique.* **39**, pp. 9-10.

Pesin, Ja. (1992). Dynamical systems with generalized hyperbolic attractors: hyperbolic, ergodic and topological properties. *Ergod. Th. & Dymam.Sys..***12**, pp. 123-151.

Sataev, E. (1999). Ergodic properties of the Belykh map. *J.Math. Sci..* **95**, pp. 2564-2575.

Schmeling, J. (1998). A dimension formula for endomorphisms – the Belykh family. *Ergod. Th. & Dymam.Sys..*18, pp. 1283-1309.

Schmeling, J. and Troubetzkoy, S. (1998). Dimension and invertibility of hyperbolic endomorphisms with singularities. *Ergod. Th. & Dymam.Sys..*18, pp. 1257-1282.

Frequency Domain Performance Analysis of Marginally Stable LTI Systems with Saturation

R.A. van den Berg, A.Y. Pogromsky, J.E. Rooda

Department of Mechanical Engineering
Eindhoven University of Technology
The Netherlands

Abstract

In this paper we discuss the frequency domain performance analysis of a marginally stable linear time-invariant (LTI) system with saturation in the feedback loop. We present two methods, both based on the notion of convergent systems, that allow to evaluate the performance of this type of systems in the frequency domain. The first method uses simulation to evaluate performance, the second method is based on describing functions. For both methods we find sufficient conditions under which a frequency domain analysis can be performed. Both methods are practically validated on an electromechanical setup and a simulation model of this setup.

5.1 Introduction

For linear systems it is common practice to analyse the performance in the frequency domain. Such an analysis provides valuable information on how the system reacts (in terms of gain and phase) to inputs with various frequencies. That is, it provides insight in how good the system can follow a certain periodic reference signal, and how it reacts on disturbances of a certain frequency.

For nonlinear systems, a similar frequency domain analysis would be very useful as well to indicate the performance of the system. However, such a frequency domain analysis is virtually impossible to perform for nonlinear systems *in general*, due to specific nonlinear behavior, such as the existence of multiple limit solutions, non-harmonic responses to harmonic inputs, dependence on initial conditions, etcetera. Nonetheless, for *some* nonlinear systems a frequency domain analysis is possible, as will be demonstrated in this paper. Some other recent results in this field can be found for example in [Jönsson *et al.*, 2003].

In this paper we focus on the class of marginally stable LTI system with saturation in the feedback loop, and discuss conditions under which it is possible to perform a frequency domain analysis for these systems. We describe and demonstrate two different approaches, both based on the notion of convergent systems, that can be used to obtain a frequency domain analysis for these nonlinear systems, i.e. a simulation approach and a describing function approach. It is interesting to note that for these marginally stable LTI systems with saturation, it is impossible to compute a finite L2-gain between input and arbitrary output using the Circle criterion or Popov criterion, while with the two approaches presented in this paper it is possible to find even more detailed input-output results than an L2-gain.

Both approaches that we discuss are based on the notion of convergent systems. Convergent systems are, roughly speaking, a class of nonlinear systems with inputs that have a unique bounded globally stable limit solution, which is dependent on the input signal. In the past, quadratic/exponential convergence and quadratically/exponentially convergent design of asymptotically stable systems has been investigated in several publication, see e.g. [Pavlov *et al.*, 2004; Pavlov *et al.*, 2007a] and references therein. However, these (design) methods are not applicable if the system is marginally stable. It is only recently that *uniform* convergence was proven for a marginally stable system [van den Berg *et al.*, 2006; van den Berg *et al.*, 2009]. If a system is uniformly (or exponentially) convergent, this implies that for each periodic input there is a unique periodic output with the same period as the input, which in turn implies that a (nonlinear) frequency response function can be found [Pavlov *et al.*, 2007b]. In the first approach, this frequency response function is found using simulation. In the second approach the frequency response function is approximated using the ideas of the describing function method, see e.g. [Khalil, 2002; Rosenwasser, 1969]. If a harmonic input is applied to the considered LTI system with saturation, then the describing function can be used to compute a linear approximation of the system, which in turn can be evaluated in the frequency domain. Since it is also possible to compute an upper bound on the error between the linear approximation and the original nonlinear system, see [van den Berg *et al.*, 2007], an interval can be indicated within which the frequency domain performance of the nonlinear system lies. Both approaches are practically validated on an experimental setup (an electromechanical system) and a simulation model of this setup.

The outline of this paper is as follows. Section 5.2 presents the class of LTI systems with saturation that is considered throughout this paper, and the electromechanical system that is used as a validation case. Furthermore it is demonstrated by means of an example why frequency domain analysis can not be performed for nonlinear systems in general. Section 5.3 deals with the notion of convergent systems and discusses how for convergent systems the simulation approach can lead to a frequency domain performance analysis. In Section 5.4 it is discussed how the describing function approach can lead to a frequency domain performance analysis. The results in Sections 5.3 and 5.4 are validated using the electromechanical system. Finally, Section 5.5 concludes the paper.

5.2   LTI system with saturation

In this section, we first describe the class of systems that is considered throughout this paper. Then, we introduce an electromechanical system within this class of systems, that will be used as a case study to practically validate the theoretical results discussed in this paper. Finally, we show by means of an example that this system can –under certain settings– exhibit rich nonlinear dynamics, which make a frequency domain analysis virtually impossible. Based on these observations, we make some statements on the conditions that a nonlinear system should satisfy in order to allow frequency domain analysis. These statements will be elaborated in Sections 5.3 and 5.4.

5.2.1   *System description*

In this paper we consider the type of systems visualized in Figure 5.1. Here, the plant dynamics are given by

$$\dot{x}_p = A_p x_p + B_p \text{sat}(u)$$
$$y_p = C_p x_p$$

where A_p has one eigenvalue at 0 and the other eigenvalues (if any) in the open left-hand plane, i.e. the plant is marginally stable. The controller dynamics are given by

$$\dot{x}_c = A_c x_c + B_c(w - y_p) + L_{AW} (\text{sat}(u) - u)$$
$$u = C_c x_c + D_c(w - y_p)$$

in which L_{AW} is a static anti-windup gain, and the saturation function is defined as $\text{sat}(u) = \text{sign}(u) \min(1, |u|)$.

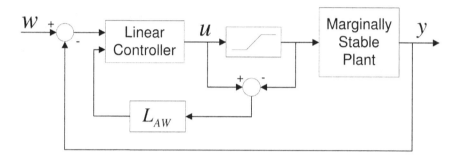

Fig. 5.1   LTI System with marginally stable plant.

The closed-loop dynamics of this system can be written in Lur'e form

$$\dot{x} = Ax + B\text{sat}(u) + Fw$$
$$u = Cx + Dw \qquad\qquad (5.1)$$
$$y = Hx$$

with state $x = [\ x_p,\ x_c\ ]^T \in \mathbb{R}^n$, input $w \in \mathbb{R}$, performance output $y \in \mathbb{R}$, matrix H to be defined freely, and

$$A = \begin{bmatrix} A_p, & 0 \\ L_{AW} D_c C_p - B_c C_p, & A_c - L_{AW} C_c \end{bmatrix},$$
$$B = \begin{bmatrix} B_p \\ L_{AW} \end{bmatrix}, \quad F = \begin{bmatrix} 0 \\ B_c - L_{AW} D_c \end{bmatrix},$$
$$C = \begin{bmatrix} -D_c C_p, & C_c \end{bmatrix}, \quad D = \begin{bmatrix} D_c \end{bmatrix}.$$

Although the theory that we present in Sections 5.3 and 5.4 applies to the whole class of systems described by (5.1), we focus in this paper on a case that has been investigated by means of simulation and real-time experiments in order to validate the theoretical findings. This case is discussed in the following subsection.

5.2.2 Case: electromechanical system

As a special case of (5.1), we consider the electromechanical system (see Figure 5.2) that is shown schematically in Figure 5.3. The hardware consists of two rotating masses connected by an element that has a certain stiffness and damping. The first mass is driven by an actuator (brushless DC motor) and the rotation of the second mass is measured by a sensor (incremental encoder, 8192 counts/revolution). The hardware is connected (at sample rate: 1 kHz) to a computer with a Matlab Simulink model (Real Time Workshop), which contains a PI controller, a saturation function and a static anti-windup gain as shown in Figure 5.3. The actuator is driven by a velocity controller (not shown in Figure 5.3), which receives its reference value v from the Simulink model. The settling time of the velocity controller is negligible, so that we can assume that the actuator exactly follows the reference velocity v.

Fig. 5.2 Case: electromechanical system (photo). From left to right: encoder, mass2, spring/damper, mass1, motor+encoder.

In order to perform also simulations on this case, the parameters of the electromechanical system have been identified and a simulation model has been created. The model is of the form (5.1) with $x_p = [r_1, \ r_2, \ \dot{r}_2]$ (r_i

denoting the rotation angle [revolutions] of mass i) and matrices

$$A = \begin{bmatrix} 0 & 0 & 0 & 0 \\ 0 & 0 & 1 & 0 \\ 3.9 \cdot 10^3 & -3.9 \cdot 10^3 & -10.7 & 0 \\ 0 & -1 + L_{AW} K_P & 0 & -L_{AW} K_I \end{bmatrix},$$

$$B = \begin{bmatrix} 1 \\ 0 \\ 10.7 \\ L_{AW} \end{bmatrix}, \quad F = \begin{bmatrix} 0 \\ 0 \\ 0 \\ 1 - L_{AW} K_P \end{bmatrix}, \tag{5.2}$$

$$C = \begin{bmatrix} 0 - K_P 0 & K_I \end{bmatrix}, \quad D = \begin{bmatrix} K_P \end{bmatrix}, \quad H = \begin{bmatrix} 0 & 1 & 0 & 0 \end{bmatrix},$$

where K_I, K_P and L_{AW} are controller parameters to be chosen. For periodic motions with relatively low frequencies and amplitudes that are neither too small or too large, the electromechanical system can be well approximated by the above model. However, for trajectories with relatively large frequencies and/or too small/too large amplitudes, the system exhibits more complex nonlinear behavior and the above model is no longer accurate.

5.2.3   *Motivating example: nonlinear behavior*

Although it is common practice for linear systems to evaluate the performance in the frequency domain, for nonlinear systems in general this is not possible. In this subsection two examples are given that clearly indicate what difficulties arise when trying to make a frequency domain analysis.

For the first example, consider the system (5.1), (5.2) with $K_I = 20$, $K_P = 8$, $L_{AW} = 0$, and $w = \sin(t)$. We evaluate the solution of this system for two initial conditions, i.e. $x(0) = [0, 0, 0, 0]$ and $x(0) = [3, 3, 0, 0]$, using both the experimental setup (see Figure 5.3) and simulation. The resulting rotation angle of mass 2 as a function of time is given in Figure 5.4.

For the second example, consider again the system (5.1), (5.2) with $K_I = 20$, $K_P = 8$, $L_{AW} = 0$, but now with $w = 5\sin(t)$. In this example we show that different initial conditions can not only lead to different 1-periodic limit solutions, but also to multi-periodic limit solutions. For four initial conditions the solution of the system is evaluated using both the experimental setup and simulation. The control output u, which clearly displays the multi-periodic solutions, is shown as a function of time in Figure 5.5.

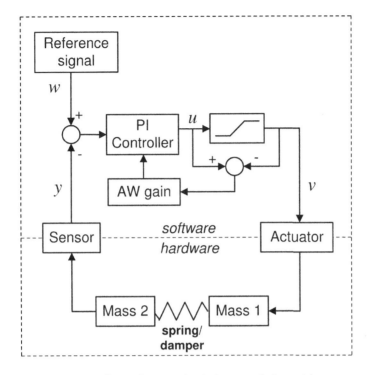

Fig. 5.3   Case: electromechanical system (schematic).

Since frequency domain analysis is based on a one-to-one mapping from input signal (e.g. reference or disturbance) to output signal (e.g. error signal or performance output) of a system, we need to guarantee that for each input signal a unique output signal exists, which has the same period as the input signal and which is independent of the initial conditions. As shown in this subsection, the output signal of a nonlinear system, however, does not necessarily satisfy these conditions, i.e. multiple 1-periodic limit solutions (see Figure 5.4) or multi-periodic limit solutions (see Figure 5.5) may exist. This motivated us to investigate whether there exist conditions under which the performance of nonlinear system (5.1) *can* be analyzed in the frequency domain.

Two approaches have been found that allow to find sufficient conditions under which a frequency domain analysis can be performed for system (5.1), i.e. a simulation-based approach and a describing function approach, which are both based on the notion of convergent systems. These approaches will be discussed in respectively Section 5.3 and Section 5.4.

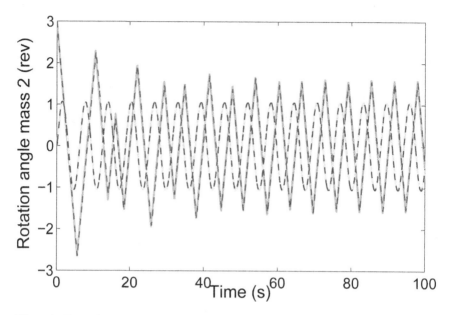

Fig. 5.4   Example 1: multiple limit solutions (experiments: dashed lines, simulations: solid lines).

5.3   Convergent systems and simulation-based frequency domain analysis

In this section, we first give a definition and some properties of convergent systems. Then, we discuss the conditions under which system (5.1) is uniformly convergent. Finally, we show how to perform a simulation-based frequency domain analysis for the convergent system.

5.3.1   Convergent systems

Consider the following class of systems,

$$\dot{x}(t) = f(x, w(t)) \tag{5.3}$$

with state $x \in \mathbb{R}^n$ and input $w \in \overline{\mathbb{PC}}$. Here, $\overline{\mathbb{PC}}$ is the class of bounded piecewise continuous inputs $w(t) : \mathbb{R} \to \mathbb{R}^m$. Furthermore, assume that $f(x, w)$ satisfies some regularity conditions to guarantee the existence of local solutions $x(t, t_0, x_0)$ of system (5.3) for any input $w \in \overline{\mathbb{PC}}$.

Definition 5.1. System (5.3) is said to be *uniformly convergent* for a class of inputs $\mathcal{W} \subset \overline{\mathbb{PC}}$ if for every input $w(t) \in \mathcal{W}$ there is a solution $\bar{x}(t) =$

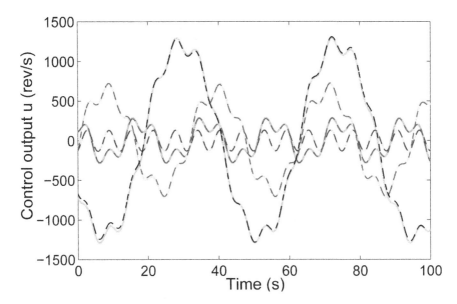

Fig. 5.5 Example 2: multi-periodic limit solutions (experiments: dashed lines, simulations: solid lines).

$x(t, t_0, \bar{x}_0)$ satisfying the following conditions:

(1) $\bar{x}(t)$ is defined and bounded for all $t \in (-\infty, +\infty)$,
(2) $\bar{x}(t)$ is globally uniformly asymptotically stable for every input $w(t) \in \mathcal{W}$.

The solution $\bar{x}(t)$ is called a *limit solution*. As follows from the above definition, any solution of an uniformly convergent system 'forgets' its initial condition and converges to a limit solution which is independent of the initial conditions. The following statements describe some properties of this limit solution.

Property 1 (Pavlov et al., 2007a). *For a uniformly convergent system, the limit solution $\bar{x}(t)$ is unique, i.e. it is the only solution bounded for all $t \in (-\infty, +\infty)$.*

Property 2 (Pavlov et al., 2006). *Suppose system (5.3) is uniformly convergent. Then, if the input $w(t)$ is constant, the corresponding limit solution $\bar{x}(t)$ is also constant. If the input $w(t)$ is periodic with period T, then the corresponding limit solution $\bar{x}(t)$ is also periodic with the same period T.*

Finally note that a system is called *exponentially convergent* for a class of inputs $\mathcal{W} \subset \overline{\mathbb{PC}}$ if it is uniformly convergent and $\bar{x}(t)$ is globally exponentially stable for every input $w(t) \in \mathcal{W}$.

5.3.2 Convergent system design

Consider again the system in Figure 5.1 with the marginally stable plant with one zero eigenvalue, as described by (5.1). Theorem 5.1 provides conditions under which this system is uniformly convergent.

Theorem 5.1. *Assume that a Lyapunov matrix $P = P^T > 0$ exists for which:*

$$PA + A^T P \le 0$$

and

$$P(A + BC) + (A + BC)^T P < 0.$$

Then, system (5.1) is uniformly convergent for all $w \in \mathcal{W}$, where $\mathcal{W}$ is the class of uniformly continuous bounded inputs. Furthermore, for any compact set Ω, if the initial condition $x(0) \in \Omega$ then the system has an exponential convergence rate for all $w \in \mathcal{W}$.

Proof.   See [van den Berg *et al.*, 2009]. $\square$

Note that if there exists a Lyapunov matrix $P = P^T > 0$ such that $PA + A^T P < 0$ and $P(A + BC) + (A + BC)^T P < 0$ hold (instead of condition 3), then the corresponding system can be proven to be exponentially convergent for all $w \in \overline{\mathbb{PC}}$, and conditions 1 and 2 of Theorem 5.1 and uniform continuity of w are not even required. However, the system we consider has a marginally stable plant thus $PA + A^T P < 0$ can not be satisfied.

5.3.3 Performance analysis in frequency domain

Under the conditions of Theorem 5.1 system (5.1) is uniformly convergent, which implies that for any input $w \in \mathcal{W}$ the system has a unique limit solution $\bar{x}$ and thus a unique output $\bar{y}$. That is, if we apply a harmonic input signal with period T to the system, then the limit output $\bar{y}$ is unique (i.e. independent of initial conditions) and has period T. Thus, we can find a kind of frequency response function if we evaluate the input-output behavior for a range of frequencies. Since the output signal is not necessarily harmonic, however, we can not obtain a typical gain and phase plot (Bode plot) as for linear systems. Instead, we determine a nonlinear frequency response function, as discussed in [Pavlov *et al.*, 2007b], i.e. for the convergent system we determine the gain between the RMS (root mean square) value of the input signal and the RMS value of the limit output signal. As phase is not defined for nonlinear systems, only the gain as discussed above will be considered in our frequency domain analysis.

Note that the periodic output $\bar{y}$ can easily be determined using *simulation* (or real-time experiments). Since the limit solution of a convergent system only depends on the input and is independent of the initial conditions, a single simulation run (experiment) suffices to determine the limit solution of the system. This is a major difference with 'non-convergent' nonlinear systems, for which in principle all initial conditions should be evaluated (i.e. an infinite amount of simulations) to obtain a reliable analysis.

This simulation-based frequency domain analysis is now demonstrated for system (5.1), (5.2). By choosing $K_I = 20$, $K_P = 8$, and $L_{AW} = 0.5$ all conditions of Theorem 5.1 are met, and thus the system is uniformly convergent for all inputs $w \in \mathcal{W}$. Evaluating the solution for the inputs signals $w = b\sin(\omega t)$ for $b = 1$ and $\omega \in [10^{-1}, 10^2]$, and computing the 'complementary sensitivity' gain ($RMS_{\bar{y}(t)}$ / $RMS_{w(t)}$) results in the frequency response function shown in Figure 5.6. Any other desired frequency response function can be computed in a similar way.

For $\omega > 10$ rad/s, the experiments and simulation give different results. Due to the relatively high frequency in combination with the saturation function, the amplitude of the motion of the masses becomes so small that nonlinear behavior of the experimental setup becomes significant, which in turn results in a different RMS-gain. However, since dealing with the undesired nonlinear behavior of the experimental setup lies outside the scope of this paper, it will not be discussed further here. In the remainder of this paper, we will focus on the dynamics as described by the simulation model.

Figure 5.6 provides valuable information on the frequency domain performance of the system. It clearly shows how the the limit output $\bar{y}$ behaves under input signals for respectively low and high frequencies. A similar plot can be made for the 'sensitivity' gain ($RMS_{w(t)-\bar{y}(t)}$ / $RMS_{w(t)}$) to investigate for example tracking behavior.

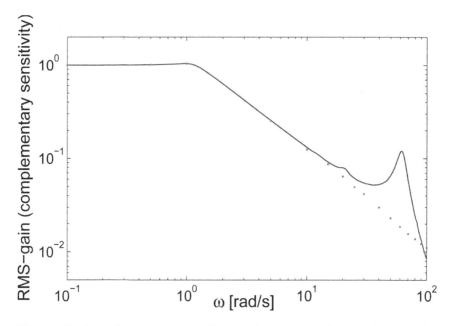

Fig. 5.6 Nonlinear frequency response function (experiments: dots, simulations: solid line).

Note, however, that the computed frequency response function in Figure 5.6 is only valid for harmonic input signals with amplitude $b = 1$. For other input amplitudes the frequency response function can be computed as well, but will be different since the limit solution $\bar{y}$ does not only depend on frequency ω but also on amplitude b. For the same reason, the superposition principle does not hold. On the other hand, computing the frequency response function for any multi-harmonic input signal is as simple as computing this function for a harmonic input signal, so the frequency response to any periodic input can be obtained by this approach.

Furthermore, note that even if we were able to find a finite L2-gain for this marginally stable system, this would only be a horizontal line in Figure 5.6, i.e. an upper bound for the frequency domain performance. Our approach based on convergence and simulation provides more detailed information on the frequency domain behavior of the system.

Although this approach leads to an exact performance analysis in the frequency domain, it can be very time-consuming, since the limit solution $\bar{x}(t)$ has to be obtained by simulation (or a real-time experiment): transient behavior should be ruled out by simulating long enough, and the simulation should be performed with high accuracy.

In the following section we will consider another approach, based on the describing function method, which is much more time-efficient, but at the cost of accuracy, i.e. instead of an exact solution, an upper- and lower bound on the performance are given. Also, this approach can only be used for harmonic input signals.

5.4   Frequency domain analysis based on describing function approach

In this section, we first give a short overview of the describing function method and explain how this theory can be expanded for application to convergent nonlinear systems with harmonic inputs. Then, we discuss a theorem which gives sufficient conditions for computation of a linear approximation and upper- and lower bound of the error of this approximation. Finally, we apply the theory on the system (5.1), (5.2).

5.4.1   *Describing function method*

Following the describing function method, the solution $\bar{x}$ of system (5.1) is approximated by a periodic limit solution $\bar{\xi}$ of the linear system

$$\begin{cases} \dot{\xi} = A\xi + BK\zeta + Fw \\ \zeta = C\xi + Dw \\ \eta = H\xi \end{cases} \tag{5.4}$$

in which gain K is to be determined. If the matrix $A + BKC$ does not have eigenvalues on the imaginary $\pm i\omega$ axis then for a periodic input $w(t) = b\sin(\omega t)$ the system has a unique periodic limit solution $\bar{\xi}(t)$, and thus a unique periodic limit output $\bar{\zeta}(t)$, which can be described by

$$\bar{\zeta}(t) = a\sin(\omega t + \psi), \tag{5.5}$$

for some amplitude $a > 0$ and phase ψ. Let $\bar{\zeta}(t)$ be given, according to the process of harmonic linearization gain K is chosen as the first Fourier coefficient in the corresponding Fourier series of $\phi(\bar{\zeta}(t))$ divided by a. Applying the fact that the saturation nonlinearity is an odd function and filling in (5.5), this simplifies to

$$K(a) = \frac{1}{\pi a} \int_0^{2\pi} \text{sat}(a\sin\theta)\sin\theta d\theta,$$

which leads to the describing function:

$$K(a) = \begin{cases} 1, & a \leq 1 \\ \frac{2}{\pi}\left(\sin^{-1}\left(\frac{1}{a}\right) + \frac{1}{a}\sqrt{1 - \frac{1}{a^2}}\right), & a > 1 \end{cases}$$

Under the assumption that A does not have eigenvalues $\pm i\omega$, the value of amplitude a can be determined by solving the so-called harmonic balance equation, which for system (5.4) is given by

$$|1 - K(a)G(i\omega)|^2 a^2 = |C(i\omega I_n - A)^{-1}F + D|^2 b^2 \qquad (5.6)$$

with $G(i\omega) = C(i\omega I_n - A)^{-1}B$. Note that the left-hand side of (5.6) is a nonlinear function of a, and the value of the right-hand side of (5.6) depends on the input amplitude b and input frequency ω. Therefore, if we want to solve this equation for a, there may exist multiple solutions of a for one pair of (b,ω), see e.g. Figure 5.7. In this Figure we plotted the left-hand side of (5.6) as a function of a, and if for example $|1 - K(a)G(i\omega)|^2 a^2 = |C(i\omega I_n - A)^{-1}F + D|^2 b^2 = 300$ for some pair of (b,ω), then multiple solution of a exist. If, on the other hand, there *is* a unique positive real solution $a(b, \omega)$ for a given pair of (b,ω), we can easily compute the limit solution $\bar{\xi}(t)$ of (5.4) by filling in $K(a(b,\omega))$, and compute how accurate this solution $\bar{\xi}(t)$ approximates $\bar{x}(t)$. However, if the solution $a(b, \omega)$ is not unique positive and real, e.g. there are multiple solutions for a, then this approach is not applicable.

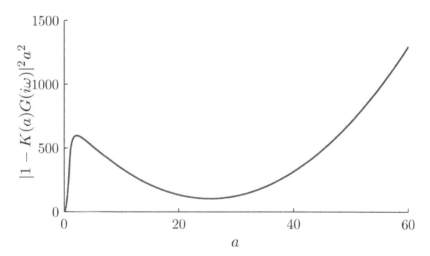

Fig. 5.7   Left hand side of (5.6) for system (5.1), (5.2) with $K_I = 20$, $K_P = 8$, $L_{AW} = 0$, $\omega = 1$.

Denote

$$\rho_1 = \sup_{k=\pm3,\pm5,\dots} |C(ik\omega I - A - \frac{\mu}{2}BC)^{-1}B|,$$

$$\bar{\rho}_1 = \sup_{k=\pm1,\pm3,\dots} |C(ik\omega I - A - \frac{\mu}{2}BC)^{-1}B|,$$

$$\rho_2 = \sup_{k=\pm3,\pm5,\dots} |H(ik\omega I - A - \frac{\mu}{2}BC)^{-1}B|,$$

$$\bar{\rho}_2 = \sup_{k=\pm1,\pm3,\dots} |H(ik\omega I - A - \frac{\mu}{2}BC)^{-1}B|$$

$$\gamma = \frac{\mu\rho_1\bar{\rho}_2}{2 - \mu\bar{\rho}_1} + \rho_2$$

$$v(a) = \left(\frac{1}{2\pi}\int_0^{2\pi}\left[\frac{2}{\pi}\int_0^{\pi}\text{sat}(a\sin\theta)\sin\theta d\theta \cdot \sin\vartheta\right.\right.$$
$$\left.\left. -\text{sat}(a\sin\vartheta)\right]^2 d\vartheta\right)^{\frac{1}{2}}.$$

to be used in the following theorem.

Theorem 5.2. *Consider system* (5.1) *with periodic input* $w(t) = b\sin(\omega t)$ *and assume the following conditions are met*

(1) (A, B) *is controllable,* (A, C) *is observable,*
(2) *matrix* A *does not have eigenvalues* $\pm i\omega$
(3) *frequency domain condition* Re $G(ik\omega) < 1/\mu$ *is satisfied for* $k = \pm1, \pm3, \pm5, \dots$,

Then the harmonic balance equation has a unique positive real solution and any half-wave symmetric $2\pi/\omega$*-periodic solution*[1] $\bar{x}(t)$ *satisfies the following relation*

$$\left(\frac{\omega}{2\pi}\int_0^{2\pi/\omega}[\bar{z}(t) - \bar{\eta}(t)]^2 dt\right)^{\frac{1}{2}} \leq \gamma v(a(b,\omega)). \tag{5.7}$$

Proof.   See [Pogromsky *et al.*, 2009].   □

---

[1] A $2\pi/\omega$-periodic function x is called half-wave symmetric if it satisfies $x(\omega t + \pi) = -x(\omega t)$

5.4.2 *Performance analysis example*

To demonstrate the use of Theorem 5.2 consider again system (5.1), (5.2), with $K_I = 20$, $K_P = 8$, $L_{AW} = 0.5$, and $w(t) = b\sin(\omega t)$ with $b = 1$ and $\omega \in [10^{-1}, 10^2]$. From Section 5.3 we know that this system is convergent. Instead of performing many time-consuming simulations, we now simply compute the linearization and error bounds for the given range of frequencies using the approach given in Subsection 5.4.1. The result is given in Figure 5.8. For comparison, the results of the simulation approach, and the gain of the linear system, i.e. system (5.1) with $\mathrm{sat}(u) = u$, are plotted as well.

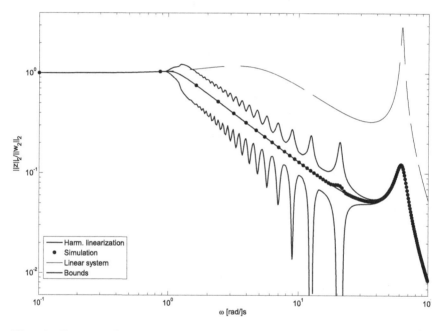

Fig. 5.8   Frequency domain results of describing function approach in comparison with simulation approach and linear system.

As one can see, the exact results as obtained with the simulation approach lie well within the error bounds of the approximation obtained by the describing function approach. Since the error bounds are relatively small for this case, the describing function approach gives a quite accurate description of the frequency domain behavior of nonlinear system (5.1), (5.2). It can also be clearly seen that the frequency domain behavior of the system with saturation substantially differs from the system without saturation.

5.5 Conclusion

Two approaches have been described that can be used to obtain a frequency domain performance analysis for a class of marginally stable LTI systems with saturation. The first step in both approaches is to prove/obtain convergence of the system. Then, the simulation approach leads to an exact performance analysis, but can be time-consuming. The other approach, based on the describing function method, is computationally much faster, but at the cost of some accuracy: only an upper- and lower bound can given on the performance of the nonlinear system, although these bounds can be very close. An electromechanical system has been used as a case to demonstrate and practically validate both approaches.

Acknowledgments

This work was partially supported by the Dutch-Russian program "DyCo-HyMS" (NWO grant 047.017.018).

References

Jönsson, U.T., Kao, C.Y. and Megretski, A. (2003). Analysis of periodically forced uncertain feedback systems, *IEEE Transactions on circuits and systems–I: fundamental theory and applications* **50**, 2, pp. 244–258.

Khalil, H.K. (2002). *Nonlinear systems*, Prentice Hall, New Jersey, third edition.

Pavlov, A., Pogromsky, A., v.d.Wouw, N. and Nijmeijer, H. (2004). Convergent dynamics, a tribute to Boris Pavlovich Demidovich, *Systems and Control Letters* **52**, pp. 257–261 .

Pavlov, A., v.d. Wouw, N. and Nijmeijer, H. (2006). *Uniform Output Regulation of Nonlinear Systems: a Convergent Dynamics Approach*, Birkhauser, ISBN 0-8176-4445-8.

Pavlov, A., Pogromsky, A., v.d.Wouw, N. and Nijmeijer, H. (2007). On convergence properties of piecewise affine systems, *International Journal of Control* **80**, 8, 1233–1247.

Pavlov, A., v.d.Wouw, N. and Nijmeijer, H. (2007). Frequency response functions for nonlinear convergent systems, *IEEE Transactions Automatic Control*, **52**, 6, pp. 1159–1165.

Pogromsky A., van den Berg, R.A. and Rooda, J.E. (2009). Frequency domain performance analysis of Lur'e systems, submitted.

Rosenwasser E.N. (1969). *Oscillations of Nonlinear Systems*, Nauka, Moscow, (in Russian).

van den Berg, R.A., Pogromsky, A. and Rooda, J.E. (2006). Convergent Systems Design: Anti-Windup for Marginally Stable Plants, *Proceedings of the 45th IEEE Conference on Decision and Control*, San Diego, USA

van den Berg, R.A., Pogromsky, A. and Rooda, J.E. (2007). Well-posedness and Accuracy of Harmonic Linearization for Lur'e Systems, *Proceedings of the 46th IEEE Conference on Decision and Control, New Orleans, USA.*

van den Berg, R.A., Pogromsky, A. and Rooda, J.E. (2009). Uniform Convergency for Anti-Windup Systems with a Marginally Stable Plant, submitted.

Reduction of Steady-State Vibrations in a Piecewise Linear Beam System Using Proportional and Derivative Control

R.H.B. Fey*, R.M.T. Wouters[†], H. Nijmeijer*

*Eindhoven University of Technology
Department of Mechanical Engineering
PO Box 513, 5600 MB Eindhoven, The Netherlands

[†]Yacht BV
Department of Technology
PO Box 12610, 1100 AP Amsterdam-Zuidoost, The Netherlands

Abstract

Control based on Proportional and/or Derivative feedback (PD control) is successfully applied to a piecewise linear beam system in order to reduce steady-state vibration amplitudes. Two control objectives are formulated: 1) to minimize the transversal vibration amplitude of the midpoint of the beam at the frequency where the first harmonic resonance occurs, and 2) to achieve this in a larger excitation frequency range. The vibration reduction that is achieved in simulations and validated by experiments is very significant for both objectives. Current results obtained with active PD control are compared with earlier results obtained using a passive linear Dynamic Vibration Absorber.

6.1 Introduction

Currently, demands concerning performance and accuracy of machinery and structures are set at high levels. Due to this the mitigation of vibrations in machinery, structures, and systems has become more important. To achieve this, passive control [den Hartog, 1985; Hunt, 1979; Korenev and Reznikov, 1993; Mead, 1999], semi-active control [Preumont, 2002; Preumont and Seto, 2008], or active control [Meirovitch, 1990; Gawronski, 2004; Preumont and Seto, 2008] can be used. The references mentioned above focus on linear systems with linear control.

Piecewise linear systems are frequently met in engineering practice, see for example [Fey and van Liempt, 2002]. Steady-state dynamics of uncontrolled piecewise linear single-dof systems and multi-dof beam systems were studied in [Fey *et al.*, 1996; Shaw and Holmes, 1983; van de Vorst *et al.*, 1996].

The objective of this chapter[1] is to investigate, to what extent a relatively simple Proportional and/or Derivative active feedback controller (a linear PD controller) can be applied, to reduce steady-state vibrations or rather, resonances, in a harmonically excited, piecewise linear beam system. In this system a one-sided spring flushes to the linear beam in the static equilibrium position. In [Bonsel *et al.*, 2004] a linear Dynamic Vibration Absorber (DVA) was used in order to passively reduce the steady-state vibrations in the same piecewise linear beam system. Obviously, application of more advanced active (nonlinear) controllers and observers [Doris, 2007] may result in increased vibration reduction, but also in increased costs, increased complexity, and lower reliability.

In Section 6.2 first the experimental setup of the piecewise linear beam system will be introduced. The steady-state behavior of the uncontrolled system will be shown in Section 6.3. Subsequently, two control objectives and the PD controller design approach will be presented in Section 6.4. Section 6.5 will discuss the numerical model of the system. In Section 6.6 first the separate effect of Proportional feedback and secondly the separate effect of Derivative feedback on the steady-state behavior of the closed loop system will be investigated. Based on the insights obtained, in Section 6.7 two PD control settings will be determined in order to realize the two control objectives. Experimental and numerical results will be compared. In Section 6.8 a brief comparison between the results obtained in this chapter and the results obtained in [Bonsel *et al.*, 2004] using a linear DVA will be carried out. Finally, in Section 6.9 conclusions will be drawn.

---

[1]This chapter is an extended version of [Fey *et al.*, 2008].

6.2   Experimental set-up

Figure 6.1 shows the schematic representation of the experimental setup of the system. The setup exists of a steel beam **a**, which is supported at each end by a leaf spring **b**.

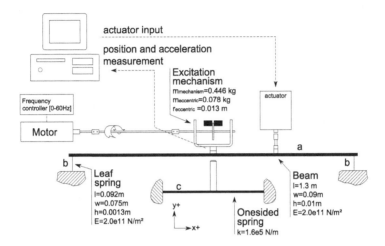

Fig. 6.1   Schematic representation of the experimental set-up.

In the middle of beam **a** a one-sided leaf spring **c** is placed. Spring **c** flushes to beam **a** (the backlash visible in Figure 6.1 is not present in reality) and makes the system piecewise linear. Spring **c** only makes contact with the main beam for downward deflection of the midpoint of beam **a**. The amount of nonlinearity in piecewise linear systems is indicated by the ratio of the one-sided stiffness and the (local) linear stiffness (of beam **a** in this case). Here, this ratio equals 2.7.

The system is transversally excited by a (disturbing) harmonic force generated by an eccentrically rotating mass mechanism, which is attached to the middle of beam **a** and is driven by an electric synchronous motor. The excitation frequency $f = \omega/2\pi$ can be varied between 0 and 60 Hz.

The actuator, which exerts the PD control force to the system, is placed as near as possible (0.2 m) to the midpoint of the beam. The operation of this actuator is based on the principle that a force is generated when a current flows through a coil, which is placed in a permanent magnetic field. The midpoint transversal displacement and acceleration are measured. The desired PD control force is determined by adding the product

of the measured displacement and a Proportional gain to the product of
the measured velocity (obtained by integrating and filtering the measured
acceleration signal) and a Derivative gain. The measurement signals are
processed by a data acquisition system, which determines an appropriate
current amplifier input for the digital PD controller to realize the desired
control force.

6.3   Steady-state behavior of the uncontrolled system

In Figure 6.2 the steady-state response of the uncontrolled system is shown.
In this figure the quantity max disp of a periodic solution of the transversal
displacement of the beam midpoint $y_{mid}(t)$, defined by:

$$\text{max disp} = \max y_{mid}(t) - \min y_{mid}(t) \tag{6.1}$$

is determined for excitation frequencies ranging from 10 to 60 Hz. Note
that a value of max disp close to zero does not necessarily mean that the
overall vibration level of the beam is close to zero, because the beam may
be vibrating in a shape with a node near or in the midpoint of the beam.

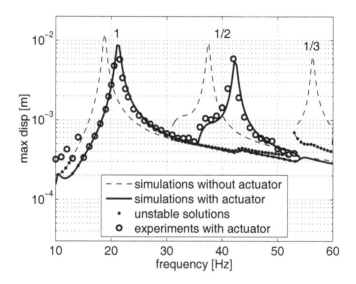

Fig. 6.2   Steady-state response of the uncontrolled system.

Figure 6.2 shows simulation results and experimental results. Simulation results are based on a numerical model, which will be introduced in

Section 6.5, and are given for the uncontrolled system *without* actuator as well as the uncontrolled system *with* passive actuator dynamics. In the former case, the stable periodic solutions are indicated with dashed lines. In the latter case, the stable periodic solutions are indicated with solid lines. In both cases unstable periodic solutions are indicated by black dots. For clarity, the uncontrolled system *with* passive actuator dynamics is the system including the inertia, damping, and elastic properties of the actuator; the actuator control force, however, is still equal to zero.

Calculation of (branches of) periodic solutions and their stability and detection of bifurcation points on these branches is based on theory and numerical methods described in e.g. [Fey *et al.*, 1996; Parker and Chua, 1989; Thomsen, 2003].

Experimental results (circles) are only included for the uncontrolled system *with* passive actuator dynamics. A good correspondence can be observed between experimental and simulation results.

Figure 6.2 shows that for both cases (*with* or *without* passive actuator dynamics) a harmonic resonance peak occurs near 20 Hz, and a related 1/2 subharmonic resonance near the double of this frequency. For the case *without* actuator also a related 1/3 subharmonic resonance is visible near 56 Hz. This resonance is also present for the case *with* passive actuator dynamics, but is not visible, since it occurs at excitation frequencies just higher than 60 Hz. So, by adding the passive actuator dynamics, the global dynamic behavior of the uncontrolled system does not change. Resonance peaks, however, shift to somewhat higher excitation frequencies.

6.4   Control objectives and PD controller design approach

From linear control theory it is known that a PD controller in principle is only capable to control a single linear normal mode. Therefore, it is important to realize that in the frequency range of interest (10-60 Hz) actually only the lowest nonlinear normal mode is dominant, see also Section 6.5. The best one may hope for is that the linear PD controller will be able to control, more or less, this single nonlinear mode. Now, two separate control objectives are formulated:

Control Objective 1 Minimize max disp defined by Equation (6.1) at the first harmonic resonance frequency of 21.2 Hz of the uncontrolled piecewise linear beam system (*with* passive actuator dynamics), see Figure 6.2.

Control Objective 2 Minimize max disp defined by Equation (6.1) in the frequency range 10-60 Hz. Here, the performance of the PD control action is evaluated by visual inspection (tuning) of multiple amplitude-frequency plots such as Figure 6.2. In this visual inspection will be focused mainly on the success of the suppression of the harmonic *and* subharmonic resonance peaks, which are present in the uncontrolled situation. Simultaneously, the appearance of new resonance peaks in the frequency range 10-60 Hz is not permitted. The visual inspection will be carried out for the whole (experimentally realizable) design space of the two PD control gains k_d and k_p, which will be introduced later.

It is emphasized again that vibration reduction of the midpoint of the beam does not guarantee *overall* vibration reduction of the beam, because for the controlled situation this midpoint may behave as a node, while the rest of the beam is still vibrating at high vibration levels. Afterwards, it will be checked if the latter situation, which obviously is undesirable with respect to overall vibration reduction, does not occur.

The beam midpoint would have been the most effective controller position, because this is the position where: 1) the disturbing harmonic excitation of the system takes place, 2) the one-sided spring is coupled to the beam, 3) the dominant 2nd eigenmode of the system without one-sided spring in the frequency range of interest shows a maximum transversal displacement (see later), and 4) the vibration reduction should be realized. However, the PD control force is applied 0.2 m from the midpoint of the beam, since from a practical point of view, for the controller this is the nearest possible position to the beam midpoint. This resembles a situation frequently occurring in engineering practice, where the control force also often cannot be applied at the ideal location.

The digital PD controller should realize the following desired PD control force F_c:

$$F_c = -k_p y_{mid} - k_d \dot{y}_{mid} \qquad (6.2)$$

where y_{mid} and $\dot{y}_{mid}$ are respectively the midpoint transversal displacement, which is directly measured, and velocity, which is obtained by integrating and filtering of the measured acceleration $\ddot{y}_{mid}$. Quantities k_p and k_d are respectively the corresponding Proportional and Derivative feedback gain.

Experimentally applicable combinations of k_p and k_d are limited because of: limited power of the amplifier-actuator combination, introduction of unstable behavior, limited accuracy of the measured signals, and limitations of the mechanical design of the actuator. This leads to the following constraints:

$$-6 \cdot 10^4 \le k_p \le 2.5 \cdot 10^4 \text{ N/m, and}$$
$$0 \le k_d \le 600 \text{ Ns/m}.$$

To realize each separate control objective, in Section 6.7 two different set-points of the PD controller will be determined. First, however, in Section 6.5 the dynamic model of the system with and without PD control will be discussed, and, to obtain more insight, in Section 6.6 the effects of separate P-action and separate D-action on the system's behavior will be investigated.

6.5 Dynamic model

A model with four degrees of freedom (dofs) is derived for efficient prediction of the dynamics of the system in the frequency range of interest (10-60 Hz):

$$M\ddot{x} + D\dot{x} + K(y_{mid})x = F \tag{6.3}$$

with:

$$x = \begin{bmatrix} y_{mid} & y_{act} & p_2 & p_3 \end{bmatrix}^T,$$

$$K(y_{mid}) = \begin{cases} K_l & \text{if } y_{mid} \ge 0, \\ K_l + K_{os} & \text{if } y_{mid} < 0, \end{cases}$$

$$F = \begin{bmatrix} m_e\omega^2 r_e cos(\omega t) & k_m G_a u & 0 & 0 \end{bmatrix}^T.$$

Matrices M, D, and K_l represent respectively the reduced mass, damping and stiffness matrix, derived by dynamic reduction of a linear Finite Element model including the main beam, the passive actuator dynamics, and the periodic excitation mechanism. The Ritz reduction matrix T, relating the reduced column x to the unreduced column $q = Tx$, is based on the 2nd eigenmode (16.2 Hz, dof p_2) and the 3rd eigenmode (54 Hz, dof p_3), see Figure 6.3, and two residual flexibility modes [Fey *et al.*, 1996], which are defined for y_{mid} and y_{act} (the transversal displacements of respectively the beam midpoint and the actuator position). These two residual flexibility modes guarantee unaffected static load behavior for the reduced model.

In the experimental set-up the 1st eigenmode is suppressed by the drive shaft of the excitation mechanism. This eigenmode is therefore excluded from the model. The 4th eigenmode and higher eigenmodes are also not taken into account in the model, since their corresponding eigenfrequencies are much higher than the frequency range of interest. It is important to realize though, that by doing so, in the frequency range of interest, possible superharmonic resonances originating from these deleted modes cannot be predicted by the reduced model.

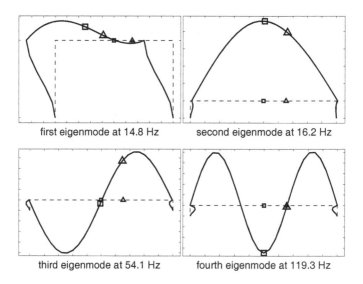

first eigenmode at 14.8 Hz   second eigenmode at 16.2 Hz

third eigenmode at 54.1 Hz   fourth eigenmode at 119.3 Hz

Fig. 6.3 First four eigenmodes (black solid lines) of undamped, unreduced linear model. Midpoint position: □, actuator position: △.

Note that the passive actuator dynamics slightly disturb the symmetry/anti-symmetry of eigenmodes with respect to the beam midpoint. Damping matrix D is assumed to account for all energy losses in the system. A choice of 2% modal damping for each mode results in a good match between measured and calculated resonance amplitudes.

The lowest eigenfrequency of the one-sided spring, see Figure 6.1, is much higher than the frequency range of interest. Therefore, the inertia of the one-sided spring may be neglected. The stiffness of the one-sided spring is represented by matrix K_{os}, containing one non-zero element at entry (1,1) equal to k_{os}.

Elements 1 and 2 of column F contain respectively the disturbing harmonic excitation force and the control force F_c. In the excitation force m_e is the rotating eccentric mass, r_e its eccentricity, and $\omega = 2\pi f$ its angular frequency. In the actuation force k_m is the motor constant, G_a the gain of the current amplifier, and u the voltage, which is chosen so, that F_c according to Equation (6.2) is realized. Obviously, the choice $u = 0$ V represents the uncontrolled situation *with* passive actuator dynamics.

The velocity signal needed for the controller was determined using analog integration of the measured acceleration signal in combination with an analog low-pass filter to reduce high-frequent noise. A model of this filter is also added to the simulation model, but detailed discussion of this filter model is considered to be out-of-the-scope of this chapter.

The nonlinear normal mode closely related to eigenmode 2 (with eigenfrequency 16.2 Hz) of the linear beam dominates the response near the first harmonic resonance peak at 21.2 Hz in Figure 6.2, and in fact in almost the whole frequency range of 10-60 Hz. The first harmonic resonance peak occurs at 21.2 Hz instead of 16.2 Hz due to the presence of the one-sided spring. The frequency shift from 16.2 Hz to 21.2 Hz can be predicted very well using the piecewise linear, single dof oscillator model derived by [Shaw and Holmes, 1983].

6.6   Effects of separate P-action and separate D-action

First, the effect of separate P-action on the response of the piecewise linear system is investigated ($k_d = 0$ Ns/m). Figure 6.4 shows max disp, see Equation (6.1), in the excitation frequency range 10-60 Hz for several values of k_p. Simulated stable periodic solutions (surfaces) are validated by experimental results (circles). Increasing k_p shifts the resonances to higher frequencies, because 'stiffness' is added to the system. The 1/2-subharmonic resonance near 42 Hz shifts approximately twice as much as the harmonic resonance near 21 Hz, because it occurs near twice the harmonic resonance frequency. For larger gains the discrepancies between numerical and experimental results increase somewhat. In Figure 6.4 only positive values of k_p are considered. Note that negative values may also be applied and in fact will be used in Section 6.7.

Figure 6.5 shows that separate Derivative feedback ($k_p = 0$ N/m) significantly suppresses both the harmonic resonance and the 1/2-subharmonic resonance near respectively 21 Hz and 42 Hz. Again the experimental and simulated stable periodic solutions match (reasonably) well.

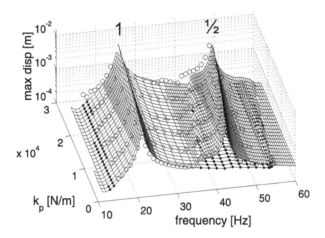

Fig. 6.4   max disp in the range 10-60 Hz for varying k_p. Surface: simulation results. Circles: experimental results. Black dots: unstable numerical results.

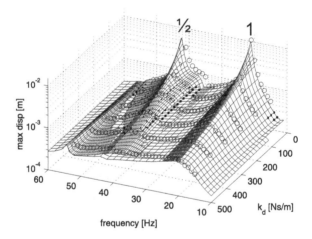

Fig. 6.5   max disp in the range 10-60 Hz for varying k_d. Surface: simulation results. Circles: experimental results. Black dots: unstable numerical results.

6.7 PD control

Now the separate effects of P- and D-action on the system behavior are known, the effect of combined P- and D-action on the systems behavior can be understood better.

6.7.1   Control objective 1

Figure 6.6 shows a contour plot of max disp, see Equation (6.1), of simulated harmonic solutions for experimentally feasible k_p, k_d combinations as defined in Section 6.4), for a constant excitation frequency of 21.2 Hz.

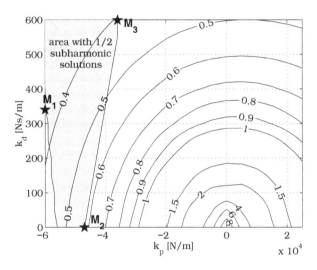

Fig. 6.6   Contour plot of max disp in [mm] in the k_p-k_d plane for an excitation frequency of 21.2 Hz.

The contours in the grey area refer to unstable harmonic solutions. Here, the stable 1/2-subharmonic resonance peak coexists, which has relatively high max disp values. Set-point $\mathbf{M_2}$ ($k_p = -4.6 \cdot 10^4$ N/m, $k_d = 0$ Ns/m) is chosen to fulfill control objective 1, because then neither k_p nor k_d needs to be set to its limit value, although a slightly higher max disp value results than in set-points $\mathbf{M_1}$ and $\mathbf{M_3}$. This value, however, still is a factor 20 lower compared to the uncontrolled maximum displacement of 10 mm (when $k_p = 0$ and $k_d = 0$).

Figure 6.7 shows the results for the controller settings of set-point $\mathbf{M_2}$ in a wider frequency range. The controlled response shows that, due to the negative Proportional feedback, the first harmonic resonance and corresponding 1/2-subharmonic resonance are shifted to lower frequencies. A local minimum of max disp, see Equation (6.1), of *stable* periodic solutions

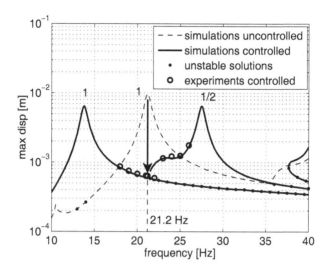

Fig. 6.7   Vibration reduction of harmonic resonance at 21.2 Hz for set-point M_2. Circles: experimental results. Black dots: unstable numerical results.

is now found at the period doubling bifurcation point at 21.2 Hz, between the two shifted resonance frequencies. Actually, at the bifurcation point itself the periodic solution is marginally stable.

It may be noted that application of negative Proportional feedback is quite unusual, since in a global sense it leads to increasing the flexibility of the structure. However, obviously, negative P-action may also shift a local minimum in the max disp-frequency plot (here corresponding to the period doubling bifurcation point) to a specific excitation frequency of interest (in this case 21.2 Hz).

For set-point $\mathbf{M_2}$ at 21.2 Hz the transversal vibration level at other locations on the beam is of the same order of, and mostly even lower than the transversal vibration level at the beam midpoint.

6.7.2   Control objective 2

When the results of various combinations of P- and D-action are visually compared (these results are not presented here), it appears that actually only D-action is needed to decrease the harmonic and subharmonic resonance peaks in order to maximize overall displacement reduction in the frequency range 10-60 Hz.

Very large P-action could shift all resonances to frequencies above 60

Hz, but cannot be applied experimentally, because amplified measurement noise results in an unstable system.

The circles in Figure 6.8 show the experimental results for the maximum experimentally applicable D-action ($k_p = 0$ N/m, $k_d = 600$ Ns/m). Both the harmonic and the 1/2-subharmonic resonance peak are reduced. The largest vibration reduction of again a factor 20 is observed at the harmonic resonance peak.

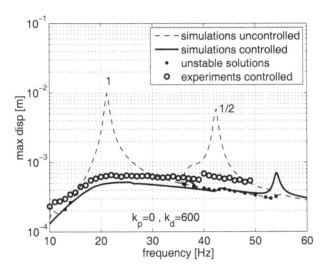

Fig. 6.8   Vibration reduction in the range 10-60 Hz for set-point $k_p = 0$ N/m, $k_d = 600$ Ns/m. Black dots: unstable numerical results of uncontrolled case.

In Figure 6.8 the difference between the simulation results for the controlled case (solid line) and the experimental results are largely due to the noise in the 'measured' velocity signal (actually the integrated and filtered measured transversal acceleration of the beam midpoint), resulting in a limited accuracy of the applied control force. Better correspondence between experimental and simulation results is obtained for lower D-action at the cost of larger max disp values.

By visual inspection it is verified that also on other locations of the beam reduction of the transversal vibrations is achieved, except for excitation frequencies near the second harmonic resonance peak, i.e. near 54 Hz. Actually, in a very small frequency range near 54 Hz, the transversal vibration level at the beam midpoint increases somewhat for the controlled situation compared to the uncontrolled situation, as can be seen in Fig-

ure 6.8. This can be understood when realizing that the control force increases the excitation of the second nonlinear normal mode, which is closely
related to the 3rd linear eigenmode at 54 Hz, as shown in Figure 6.3. It is
only due to the passive dynamics of the PD controller in the uncontrolled
situation that this eigenmode has a node, which not exactly coincides with
the beam midpoint. In fact, near 54 Hz, the transversal vibration levels
at some beam locations will be even higher than the transversal vibration
level at the beam midpoint presented in Figure 6.8. This is understandable
by examining the shape of the 3rd linear eigenmode (or rather, the shape
of the corresponding nonlinear mode).

6.8   Comparison with passive control via a linear DVA

Passive vibration control of the same system was applied in [Bonsel et al.,
2004] by fixing a linear Dynamic Vibration Absorber (DVA), see Figure 6.9,
to the beam. This DVA was fixed to the beam at the same position as the
PD-controller. In its operational mode the DVA is basically a single dof
mass-damper-spring system; in Figure 6.9 both masses at the ends of the
leaf springs will vibrate in-phase. The eigenfrequency of the DVA was tuned
to the first harmonic resonance frequency of the piecewise linear system.
In [Bonsel et al., 2004] the same two control objectives were formulated as
in this chapter. The *undamped* DVA was applied to minimize the vibration
amplitude at the first harmonic resonance (control objective 1), whereas the
damped DVA was used to maximize vibration reduction in the frequency
range 10-60 Hz (control objective 2).

The amplitudes of periodic solutions obtained by application of a DVA
are shown in Figure 6.10 for control objective 1 and in Figure 6.11 for
control objective 2.

With respect to control objective 1, it can be seen that the undamped
DVA (Figure 6.10) realizes a larger vibration reduction at the harmonic
resonance frequency than P-control (Figure 6.7). By tuning the eigenfrequency of the undamped DVA to the first harmonic resonance frequency of
the original system, two harmonic resonance peaks result on both sides of
the original resonance peak with in between an anti-resonance realizing an
enormous vibration reduction.

With respect to control objective 2, it is clear that D-control (Figure 6.8)
gives lower max disp values than the damped DVA (Figure 6.11), except for
excitation frequencies near 54 Hz. The experimental results of the system

Fig. 6.9   The undamped Dynamic Vibration Absorber.

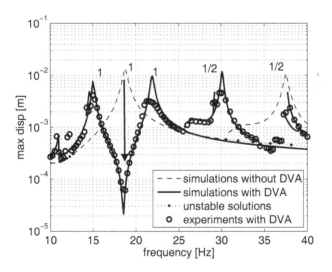

Fig. 6.10   Suppression of the first harmonic resonance peak near 19 Hz using an undamped DVA.

with the damped DVA differ substantially from the simulation results due to the fact, that the maximum value of the damping coefficient, which could be realized in the experiment, was lower than the optimal numerical value for the DVA damper characteristic. In spite of this, the vibration reduction realized with active D-control still is larger than the simulated results with the optimally damped DVA.

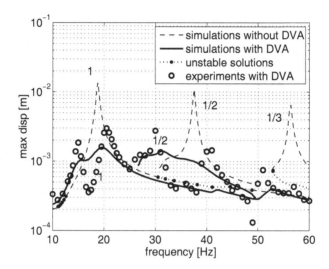

Fig. 6.11   Steady-state vibration reduction in the frequency range 10-60 Hz using a damped DVA.

6.9   Conclusions

It has been shown that a linear PD controller can be effectively used for reduction of steady-state vibrations of harmonically excited, piecewise linear systems, i.e. for the considered case of flush with a moderate amount of nonlinearity. Simulation results have been validated by experimental results and a (reasonably) good correspondence has been obtained.

First, the effects of P-action and D-action have been investigated separately. Proportional feedback mainly shifts the excitation frequencies where (sub)harmonic resonances occur. Application of D-action results in substantial reduction of the resonance amplitudes in the frequency range of interest (10-60 Hz).

Subsequently, various experimentally feasible combinations of Proportional and Derivative action have been investigated, to determine which combinations fulfill the two control objectives as good as possible. It appears that pure (negative) P-action gives the best result for the first control objective, whereas pure D-action gives the best result for the second control objective. For the case of flush considered here, the results indicate that the effect of PD control applied to a piecewise linear system is to a large extent comparable to the effect of PD control applied to a linear system.

The results obtained with the active PD controller and the results obtained with the passive DVA have been compared. With respect to the first control objective the undamped DVA results in a lower response amplitude than active Proportional control. With respect to the second control objective, in the frequency range of interest, active Derivative control on average leads to lower response amplitudes than the damped DVA. However, it must be noted that obviously, the passive DVA and the active PD controller are different devices with different properties; for instance, they have a different mass and different power capacities. This hampers a fair comparison between the performances of the passive and the PD controller (for each of the two control objectives).

As stated before, the vibration reduction realized for the transversal displacement of the midpoint of the beam does not guarantee vibration reduction for transversal displacements of other positions on the beam. However, by visual inspection it was verified, that also on other beam positions vibration reduction was achieved, except for excitation frequencies near the second harmonic resonance peak, i.e. near 54 Hz.

Acknowledgments

This work was partially supported by the Dutch-Russian program on interdisciplinary mathematics Dynamics and Control of Hybrid Mechanical Systems (NWO grant 047.017.018) and the HYCON Network of Excellence (contract no. FP6-IST-511368).

References

Bonsel, J.H., Fey, R.H.B. and Nijmeijer H. (2004). Application of a dynamic vibration absorber to a piecewise linear system, *Nonlinear Dynamics*, **37**(3), pp. 227-243.

Doris, A. (2007). Output-feedback design for non-smooth mechanical systems: control synthesis and experiments, PhD thesis, Eindhoven University of Technology, 210p.

Fey, R.H.B., Campen, D.H. van and Kraker, A. de (1996). Long term structural dynamics of mechanical systems with local nonlinearities, *ASME Journal of Vibration and Acoustics*, **118**(2), pp. 147-153.

Fey, R.H.B. and Liempt, F.P.H. van (2002). Sine sweep and steady-state response of simplified solar array models with nonlinear elements, *Proc. Int. Conf.*

on *Structural Dynamics Modelling*, Funchal, Madeira, Portugal, June 3-5, 2002, pp. 201-210.

Fey, R.H.B., Nijmeijer, H., and Wouters, R.M.T. (2008). Steady-state vibration mitigation in a piecewise linear beam system using PD control, *Proc. 6th EUROMECH Nonlinear Dynamics Conference (ENOC 2008)*, St. Petersburg, Russian Federation, June 30-July 4, 2008.

Gawronski, W.K. (2004). *Advanced Structural Dynamics and Active Control of Structures*, Springer Verlag, New York.

Hartog, J.P. den (1985). *Mechanical Vibrations*, Dover Publications, 4th edition.

Hunt, J.B. (1979). *Dynamic Vibration Absorbers*, Mechanical Engineering Publications, London.

Korenev, B.G. and Reznikov, L.M. (1993). *Dynamic Vibration Absorbers: Theory and Technical Applications*, John Wiley & Sons, Chichester.

Mead, D.J. (1999). *Passive Vibration Control*, Wiley & Sons, Chichester.

Meirovitch, L. (1990). *Dynamics and Control of Structures*, John Wiley, New York.

Parker, T.S. and Chua, L.O. (1989). *Practical Numerical Algorithms for Chaotic Systems*, Springer-Verlag, Berlin.

Preumont, A. (2002). *Vibration Control of Active Structures - An Introduction*, Kluwer Academic Publishers, Dordrecht, the Netherlands, 2nd edition.

Preumont, A. and Seto, K. (2008). *Active Control of Structures*, John Wiley & Sons, Chichester, UK.

Shaw, S.W. and Holmes, P.J. (1983). A periodically forced piecewise linear oscillator, *Journal of Sound and Vibration*, **90**(1), pp. 129-155.

Thomsen, J.J. (2003). *Vibrations and Stability, Advanced Theory, Analysis, and Tools*, Springer, Berlin, 2nd edition.

Vorst, E.L.B. van de, Assinck, F.H., Kraker, A. de, Fey, R.H.B. and Campen, D.H. van (1996). Experimental verification of the steady-state behaviour of a beam system with discontinuous support, *Experimental Mechanics*, **36**(2), pp. 159-165.

Hybrid Quantised Observer for Multi-input-multi-output Nonlinear Systems

A.L. Fradkov*, B.R. Andrievskiy*, R.J. Evans[†]

*Institute for Problems of Mechanical Engineering, RAS,
St. Petersburg, 61, VO, Bolshoy,
Russia

[†]Department of Electrical and Electronic Engineering,
University of Melbourne, Parkville Victoria 3010,
Australia

Abstract

The problem of state estimation with limited information capacity of the coupling channel for oscillatory multi-input-multi-output nonlinear systems is analyzed for systems in Lurie form (linear part plus nonlinearity depending only on measurable outputs) with first-order coder-decoder. It is shown that for oscillatory systems the upper bound of the limit estimation error is proportional to the maximum rate of the coupling signal and inversely proportional to the information transmission rate (channel capacity). The results are applied to state estimation of a 3rd order nonlinear system: self-excited mechanical oscillator.

7.1 Introduction

During the last decade substantial interest has been shown in *networked control systems* (NCS). The idea is to use serial communication networks to exchange system information and control signals between various physical components of the systems that may be physically distributed. NCS are real-time systems where sensor and actuator data are transmitted through shared or switched communication networks, see e.g. [Ishii and Francis (2002); Goodwin *et al.* (2004); Abdallah and Tanner (2007); Matveev and Savkin (2008)]. Transmitting sensor measurement and control commands over wireless links allows rapid deployment, flexible installation, fully mobile operation and prevents the cable wear and tear problem in an industrial environment. The possibility of NCS motivates development of a new chapter of control theory in which control and communication issues are integrated, and all the limitations of the communication channels are taken into account. The introduction of a communication network into a NCS can degrade overall control system performance through quantisation errors, transmission time delays and dropped measurements. The network itself is a dynamical system that exhibits characteristics which traditionally have not been taken into account in control system design. These special characteristics include quantization and time-delays and are a consequence of the fact that practical channels have limited bandwidth. A successful NCS design should take network characteristics into account. Due to the digital nature of the communication channel, every transmitted signal is quantized to a finite set [Ishii and Francis (2002)]. Hence, we argue that the finite set nature of the data should be explicitly taken into account in the design of NCS.

Recently the limitations of estimation and control under constraints imposed by a finite capacity information channel have been investigated in detail in the control theoretic literature, see [Wong and Brockett (1997); Nair and Evans (2003); Nair *et al.* (2004); Bazzi and Mitter (2005); Nair *et al.* (2007)] and references therein. For unstable linear systems it was shown that stabilisation of the system at the equilibrium under information constraints is possible if and only if the capacity of the information channel exceeds the entropy production of the system at the equilibrium (so called *Data-Rate Theorem* [Nair and Evans (2003); Nair *et al.* (2004)].

For discrete-time nonlinear systems, the concept of feedback topological entropy was introduced and the minimum data-rate for local stabilisation was given in [Nair et al. (2004)]. In [Savkin (2006)] the concept of topological entropy was extended to uncertain systems. Using this concept, in [Savkin (2006)] necessary and sufficient conditions for the robust observability of a class of uncertain nonlinear systems, and the solvability of the optimal control problem via limited capacity communication channels were obtained. Continuous-time nonlinear systems were considered in [Liberzon and Hespanha (2005); De Persis (2005); De Persis and Isidori (2004); De Persis (2006); Cheng and Savkin (2007)], where several sufficient conditions for different estimation and stabilisation problems were obtained. In [De Persis (2006)], uniformly observable systems were considered and an "embedded-observer" decoder and a controller were designed, which semi-globally stabilizes this class of systems under data-rate constraints. In [Cheng and Savkin (2007)], an output feedback stabilisation problem of a class of nonlinear systems with nonlinearities satisfying the non-decreasing property was considered, encoding/decoding scheme was introduced and the sufficient conditions for the stabilisation problem were obtained.

In most of the above mentioned papers the coding-decoding procedure is rather complicated: the size of the required memory exceeds or equals to the dimension of the system state space. Such a draw-back was overcome in [Fradkov *et al.* (2006)], where a first order coder scheme was proposed for single-input-single-output systems. Complexity of the scheme of [Fradkov *et al.* (2006)] does not grow with the dimension of the system state. In addition, in [Fradkov *et al.* (2006)] the synchronization rather than stabilization problem was studied (synchronization is irredusable to stabilization).

In this paper we extend the results of [Fradkov *et al.* (2006)] to systems with many inputs and many outputs (MIMO systems). We establish limit possibilities of state estimation for a class of nonlinear systems under information constraints. Such systems are well studied without information constraints [Morgül and Solak (1996); Nijmeijer and Mareels (1997); Nijmeijer (2001)]. Here we present a theoretical analysis for n-dimensional systems represented in the so called Lurie form (linear part plus nonlinearity, depending only on measurable outputs). It is shown that the upper bound of the limit estimation error (LSE) is proportional to the upper bound of the transmission error. As a consequence, the upper and lower bounds of LSE are proportional to the maximum rate of the coupling signal and inversely proportional to the information transmission rate (channel capacity). Optimality of the binary coding for coders with one-step memory is established.

7.2 Description of state estimation over the limited-band communication channel

A block-diagram for implementing state estimation via a discrete communication channel is shown in Fig. 7.1. To simplify exposition we will consider a system model in the so-called Lurie form: right-hand side is split into a linear part and a nonlinearity vector depends only on the measured output. Then the system is modeled as follows:

$$\dot{x}(t) = Ax(t) + \varphi\big(y(t)\big), \ y(t) = Cx(t), \tag{7.1}$$

where x is an n-dimensional (column) vector of state variables; y is an l-dimensional output vector; A is an $(n \times n)$-matrix; C is $(l \times n)$-matrix, $\varphi(y)$ is a continuous nonlinear vector-function, $\varphi : \mathbb{R}^l \to \mathbb{R}^n$. We assume that the system is dissipative: all the trajectories of the system (7.1) belong to a bounded set Ω (e.g. attractor of a chaotic system). Such an assumption is typical for oscillating and chaotic systems.

The observer has the following form:

$$\dot{\hat{x}} = A\hat{x} + \varphi(y) + L(y - \hat{y}), \ \hat{y} = C\hat{x}, \tag{7.2}$$

where L is the vector of the observer parameters (gain). Apparently, the dynamics of the state error vector $e(t) = x(t) - \hat{x}(t)$ are described by a linear equation

$$\dot{e} = A_L e, \ y = Cx, \tag{7.3}$$

where $A_L = A - LC$.

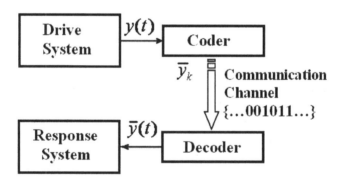

Fig. 7.1   Block-diagram for state estimation over a discrete communication channel.

As is known from control theory, if the pair (A, C) is observable, i.e. if $\text{rank}[C^\mathrm{T}, A^\mathrm{T}C^\mathrm{T}, \ldots, (A^\mathrm{T})^{n-1}C^\mathrm{T}] = n$, then there exists L providing the matrix A_L with any given eigenvalues. In particular all eigenvalues of A_L can have negative real parts, i.e. the system (7.3) can be made asymptotically stable and $e(t) \to 0$ as $t \to \infty$. Therefore, in the absence of measurement and transmission errors the estimation error decays to zero.

Now let us take into account transmission errors. We assume that the observation signal $y_i(t) \in \mathbb{R}^l$ is coded with symbols from a finite alphabet at discrete sampling time instants $t_{k,i} = k_i T_{s,i}$, where $i = 1, \ldots, l$, $k_i \in \mathbb{Z}$, $T_{s,i}$ are the sampling periods. Let the coded symbols $\bar{y}_i[k] = \bar{y}(t_{k,i})$ be transmitted over a digital communication channel with a finite capacity. To simplify the analysis, we assume that the observations are not corrupted by observation noise; transmissions delay and transmission channel distortions may be neglected. Therefore, the discrete communication channel with the sampling periods $T_{s,i}$ is considered, but it is assumed that the coded symbols are available at the receiver side at the same sampling instant $t_{k,i} = k_i T_{s,i}$, as they are generated by the coder.

Assume that *zero-order extrapolation* is used to convert the digital sequences $\bar{y}_i[k]$ to the continuous-time input of the response system $\bar{y}_i(t)$, namely, that $\bar{y}_i(t) = \bar{y}_i[k_i]$ as $k_i T_{s,i} \leq t < (k_i + 1)T_{s,i}$. Then the *transmission error vector* is defined as follows:

$$\delta_y(t) = y(t) - \bar{y}(t) \in \mathbb{R}^l. \tag{7.4}$$

In the presence of transmission errors, equation (7.3) takes the form

$$\dot{e} = A_L e + \varphi(y) - \varphi\big(y + \delta_y(t)\big) - L\delta_y(t) \tag{7.5}$$

Our goal is to evaluate limitations imposed on the estimation precision by limited transmission rate. To this end introduce an upper bound on the limit estimation error $Q = \sup \varlimsup\limits_{t\to\infty} \|e(t)\|$, where $e(t)$ is from (7.5), $\|\cdot\|$ denotes the Euclidian norm of a vector, and the supremum is taken over all admissible transmission errors. In the next two sections we describe encoding and decoding procedures and evaluate the set of admissible transmission errors $\delta_y(t)$ for the optimal choice of coder parameters. It will be shown that $\|\delta_y(t)\|$ is bounded and does not tend to zero.

The properties of state estimation over a limited-band communication channel for single-output Lurie systems with one-step memory time-varying coder are studied in [Fradkov *et al.* (2006)]. It is shown that the upper bound of the limit state estimation error is proportional to the upper bound of the transmission error. Under the assumption that a sampling time may be properly chosen, optimality of the binary coding in the sense of demanded transmission rate is established, and the relation between estimation accuracy and an optimal sampling time is found. On the basis of these results, the present paper deals with a binary coding procedure. We extend the results of [Fradkov *et al.* (2006)] for multi-output systems.

7.3   Coding procedure

We assume that the observation vector $y(t) \in \mathbb{R}^l$ is coded with symbols from a finite alphabet at discrete sampling time instants $t_{k,i} = k_i T_{s,i}$, $k_i \in \mathbb{Z}$, $i = 1, \ldots, l$, $T_{s,i}$ are sampling periods. We apply the coding/decoding procedure for each component $y_i(t)$ ($i = 1, \ldots, l$) of the system output $y(t) \in \mathbb{R}^l$ independently.

Let the coded symbols $\bar{y}_i[k_i] = \bar{y}_i(t_{k_i})$ ($i = 1, \ldots, l$, $k_i \in \mathbb{Z}$) be transmitted over a digital communication channel with finite capacity. To simplify the analysis, we assume that the observations are not corrupted by observation noise; transmissions delay and transmission channel distortions may be neglected. Therefore, the discrete communication channel with the sampling period T_s is considered, but it is assumed that the coded symbols are available at the receiver side at the same sampling instant $t_k = kT_s$, as they are generated by the coder.

At first, introduce the memoryless (static) binary coder to be a discretized map $q_\varkappa : \mathbb{R} \to \mathbb{R}$ as

$$q_\varkappa(y) = \varkappa \mathrm{sign}(y), \tag{7.6}$$

where $\mathrm{sign}(\cdot)$ is the *signum* function: $\mathrm{sign}(y) = 1$, if $y \geq 0$, $\mathrm{sign}(y) = -1$, if $y < 0$; parameter $\varkappa$ may be referred to as a *coder range* or as a *saturation value*. Evidently, $|y - q_\varkappa(y)| \leq \varkappa$ for all y such that $y : |y| \leq 2\varkappa$. Notice that for binary coder each codeword symbol contains $R = 1$ bit of information. The discretized output of the considered coder is found as $\bar{y} = q_\varkappa(y)$ and we assume that the coder and decoder make decisions based on the same information.

In the present paper we use l independent coders for components of an l-dimensional vector, transmitted over the channel. Each coder number i, $i = 1, \ldots, l$, has its particular sampling period $T_{s,i}$, transmission rate $R_i = 1/T_{s,i}$ and range $\varkappa_i$. The overall averaged rate R is calculated as a sum $R = \sum_{i=1}^{l} R_i$.

The static coder (7.6) is a part of the time-varying coders with memory, see e.g. [Nair and Evans (2003); Brockett and Liberzon (2000); Tatikonda and Mitter (2004); Fradkov *et al.* (2006)]. Two underlying ideas are used for this kind of coder:

- reducing the coder range $\varkappa$ to cover the some area around the predicted value for the $(k+1)$th observation $y[k+1]$, $y[k+1] \in \mathcal{Y}[k+1]$. This means that the quantizer range $\varkappa$ is updated during the time and a time-varying quantizer (with different values of $\varkappa$ for each instant, $\varkappa = \varkappa[k]$) is used. Using such a "zooming" strategy it is possible to increase coder accuracy in the steady-state mode, and, at the same time, to prevent coder saturation at the beginning of the process;

– introducing memory into the coder, which makes it possible to predict the $(k+1)$th observation $y[k+1]$ with some accuracy and, therefore, to transmit over the channel only encrypted innovation signal.

Let us describe the first-order (one-step memory) coder . Introduce the sequence of central vectors (sequence of "*centroids*") $c[k] \in \mathbb{R}^l$, $k \in \mathbb{Z}$ with initial condition $c[0] = 0$ [Tatikonda and Mitter (2004)]. At step k the coder compares the current measured output $y[k]$ with the number $c[k]$, forming the deviation vector $\partial y[k] = y[k] - c[k] \in \mathbb{R}^l$. Then this vector is discretized with a given $\varkappa = \varkappa[k]$ according to (7.6). The output signal

$$\bar{\partial} y[k] = q_\varkappa(\partial y[k]) \tag{7.7}$$

is represented as an R-bit information symbol from the coding alphabet and transmitted over the communication channel to the decoder. Then the central number $c[k+1]$ and the range parameter $\varkappa[k]$ are renewed based on the available information about the drive system dynamics. Assuming that the system output $y(t)$ changes at a slow rate, i.e. that $y[k+1] \approx y[k]$. we use the following update algorithms:

$$c[k+1] = c[k] + \bar{\partial} y[k], \quad c[0] = 0, \tag{7.8}$$

$$\varkappa[k] = (\varkappa_0 - \varkappa_\infty)\rho^k + \varkappa_\infty, \quad k = 0, 1, \ldots, \tag{7.9}$$

where $0 < \rho \le 1$ is the decay parameter, $\varkappa_\infty$ stands for the limit value of $\varkappa$. The initial value $\varkappa_0$ should be large enough to capture all the region of possible initial values of y_0.

The equations (7.6), (7.7), (7.9) describe the coder algorithm. A similar algorithm is used by the decoder. Namely: the sequence of $\varkappa[k]$ is reproduced at the receiver node utilizing (7.9); the values of $\bar{\partial} y[k]$ are restored with given $\varkappa[k]$ from the received codeword; the central numbers $c[k]$ are found in the decoder in accordance with (7.8). Then $\bar{y}[k]$ is found as a sum $c[k] + \bar{\partial} y[k]$.

7.4 Evaluation of state estimation error

Now let us evaluate the estimation error

$$Q = \sup \overline{\lim_{t \to \infty}} \|e(t)\|, \tag{7.10}$$

where sup is taken over the set of transmission errors $\delta_{y,i}(t)$ not exceeding the corresponding level Δ_i where $i = 1, \dots, l$. Owing to nonlinearity of the equation (7.5) evaluation of the estimation error Q is nontrivial and it may even be infinite for rapidly growing nonlinearities $\varphi(y)$. To obtain a reasonable upper bound for Q we assume that the nonlinearity is Lipschitz continuous along all the trajectories of the system (7.1). More precisely, we assume existence of some positive real number $L_\varphi > 0$ such that $\|\varphi(y) - \varphi(y+\delta)\| \le L_\varphi |\delta|$ for all $y = Cx$, $x \in \Omega$, where Ω is a set containing all the trajectories of the system (7.1), starting from the set of initial conditions Ω_0, $|\delta| \le \Delta$.

The error equation (7.5) can be represented as

$$\dot{e} = A_L e + \xi(t), \qquad (7.11)$$

where $\|\xi(t)\| \le \left(L_\varphi + \|L\|\right)\sqrt{n}\max_i \Delta_i$, i.e. the problem is reduced to a standard problem of linear system theory. Choose L such that A_L is a Hurwitz (stable) matrix and choose a positive-definite matrix $P = P^T > 0$ satisfying the modified Lyapunov inequality

$$PA_L + A_L{}^T P \le -\mu P, \qquad (7.12)$$

for some $\mu > 0$. Note that the solutions of (7.12) exist if and only if $\mu > \mu_*$, where $\mu_* = -\max \operatorname{Re}\lambda_i(A)$ is stability degree of matrix A. After simple algebra we obtain the differential inequality for the function $V(t) = e(t)^T Pe(t)$: $\dot{V} \le -\mu V + e^T P\xi(t) \le -\mu V + \sqrt{V} \cdot \sqrt{\xi^T P\xi}$. Since $\dot{V} < 0$ within the set $\sqrt{V} > \mu^{-1}\sup_t \sqrt{\xi(t)^T P\xi(t)}$, the value of $\varlimsup_{t\to\infty} V(t)$ cannot exceed $n \max_i \Delta_i^2 \left(L_\varphi + \|L\|\right)^2 \lambda_{\max}(P)/\mu^2$. In view of positivity of P, $\lambda_{\min}(P)\|e(t)\|^2 \le V(t)$, where $\lambda_{\min}(P)$, $\lambda_{\max}(P)$ are minimum and maximum eigenvalues of P, respectively. Hence

$$\varlimsup_{t\to\infty} \|e(t)\| \le C_e^+ \sqrt{l}\max_i \Delta_i, \qquad (7.13)$$

where

$$C_e^+ = \sqrt{\frac{\lambda_{\max}(P)}{\lambda_{\min}(P)}} \frac{L_\varphi + \|L\|}{\mu}. \qquad (7.14)$$

The inequality (7.13) shows that the total estimation error is proportional to the upper bound of the norm $\sqrt{n}\max_i \Delta_i$ of the transmission error. As was shown in [Fradkov et al. (2006)], a binary coder is optimal in the sense of bit-per-second rate, and the optimal sampling times $T_{s,i}$ for each channel are

$$T_{s,i} = \varepsilon \Delta_i / L_{y,i\cdot}, \qquad (7.15)$$

In (7.15), $L_{y,i}$ is the exact bound for the rate of $y_i(t)$, $L_{y,i} = \sup_{x \in \Omega} |C\dot{x}|$, where $\dot{x}$ is from (7.1), $(i = 1, \ldots, l)$; ε is the constant number, $\varepsilon \approx 0.5923$. Consequently, the channel bit-per-second rate $R = \sum_i^l T_{s,i}^{-1}$ is as follows:

$$R = \sum_i^l r L_{y,i} / \Delta_i, \tag{7.16}$$

where $r = \varepsilon^{-1} \approx 1.688$, and this bound is tight for the considered class of coders. Taking into account the relation (7.16) for optimal transmission rate, the limit state estimation error can be estimated as follows:

$$\overline{\lim_{t \to \infty}} \|e(t)\| \leq C_e^+ r \cdot l \max(L_{y,i}/R_i), \tag{7.17}$$

i.e. it can be made arbitrarily small for sufficiently large transmission rate $R = \sum_i^l R_i$.

Remark. Relations (7.15), (7.16) are related to calculation of the partial bit-rates R_i for transmission of ith component y_i of the measured vector $y(t) \in \mathbb{R}^l$ for the given partial transmission error Δ_i and the bound $L_{y,i}$ of the rate of $y_i(t)$. The problem of reallocation of the given channel bit-rate R between the components of the transmitted vector y, minimizing the estimate (7.17) of the limit estimation error may be posed. Simple calculations lead to the following optimal bit-rates R_i^*:

$$R_i^* = R \cdot L_{y,i} / \sum_i^l L_{y,i}, \quad i = 1, \ldots, l. \tag{7.18}$$

7.5 Example. State estimation of nonlinear oscillator

Let us apply the above results to state estimation of nonlinear self-excited oscillator via a channel with limited capacity.

System Equations. Consider the following self-excited nonlinear oscillator:

$$\begin{cases} \dot{x}_1 = x_2, \\ \dot{x}_2 = -\omega_1^2 \sin y_1 - \varrho x_2 + k_p \arctan(k_c y_2), \\ \dot{x}_3 = \omega_2 \cdot (x_1 - x_3), \end{cases} \tag{7.19}$$

$$y_1 = x_1, \qquad y_2 = x_1 - x_3$$

where $y(t) \in \mathbb{R}^2$ is the sensor output vector (to be transmitted over the communication channel), ω_1, ω_2, ϱ, k_p, k_c are system parameters, $x = [x_1, x_2, x_3]^\mathrm{T} \in \mathbb{R}^3$ is the state vector. The problem is to produce estimate $\hat{x}(t) \in \mathbb{R}^3$ of the system state vector $x(t)$ based on the signals $y_1(t)$, $y_2(t)$, transmitted over the communication channel.

System (7.19) has the form (7.1), where

$$A = \begin{bmatrix} 0 & 1 & 0 \\ 0 & -\varrho & 0 \\ \omega_2 & 0 & -\omega_2 \end{bmatrix}, \quad C = \begin{bmatrix} 1, 0, & 0 \\ 1, 0, & -1 \end{bmatrix},$$

$$\varphi(y) = \begin{bmatrix} 0 \\ \omega_1^2 \sin y_1 + k_p \arctan(k_c y_2) \\ 0 \end{bmatrix}. \tag{7.20}$$

For the considered case, the observer (7.2) has (3×2) gain matrix L. The matrices A, C and function $\varphi(y)$ are given in (7.20). Matrix L should be chosen so that the observer (7.2) stability conditions are satisfied, i.e. the characteristic polynomial $D_L(s) = \det(s\mathbf{I} - A_L)$ is Hurwitz. In our study the matrix L was found through the solution P of the Riccati equation as follows

$$PA^\mathrm{T} + A^\mathrm{T}P + PGP - C^\mathrm{T}C = 0, \quad L = P^{-1}C^\mathrm{T}, \tag{7.21}$$

where $G = G^\mathrm{T} > 0$ is (3×3)-matrix, such as a pair $(A, \sqrt{Q})$ is controllable. Such a choice of the gain L yields passivity of the observer (7.2) error dynamics [Shim et al. (2003)].

System parameters. For simulation the following parameter values of the oscillating system (7.19) were taken: $\omega_0^2 = 40$ s^{-2}, $\varrho = 1$ s^{-1}, $k = 0.1$, $k_p = 5$, $k_c = 10$, $\tau = 0.1$ s, $L_{y,1} = 10$ s^{-1}, $L_{y,2} = 5$ s^{-1}, the decay parameters $\rho_i = \exp(-0.2T_{s,i})$, $i = 1, 2$. The matrix G in (7.21) was taken $G = 0.1\mathbf{I}_3$, where $\mathbf{I}_3$ is an identity (3×3)-matrix. The values of $\Delta_1 = \Delta_2$ have been taken from the interval $\Delta \in \{1, 40\}$. The sampling times $T_{s,i}$ for each Δ_i weretaken from (7.15). The initial values $\varkappa_{0,i}$ in (7.9) were $\varkappa_{0,1} = 2$, $\varkappa_{0,2} = 1$.

Some simulation results for the coder (7.6), (7.7), (7.9) are shown in Figs. 7.2–7.4. The time histories of $x_1(t)$, $\hat{x}_1(t)$, estimation error $e_1(t) = x_1(t) - \hat{x}_1(t)$ for $R = 50$ bit/s, ($T_{s,1} = 0.03$ s, $T_{s,2} = 0.06$ s are shown in Fig. 7.2. Respectively, the time histories for the second state variable are depicted in Fig. 7.3. Dependence of the estimation error Q on the transmission rate $\bar{R}$ is shown in Fig. 7.4, demonstrating that the estimation error becomes small for sufficiently large transmission rates.

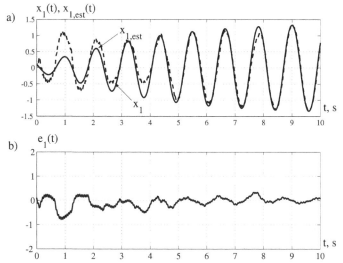

Fig. 7.2   Time histories: a) $x_1(t)$ (solid line), $\hat{x}_1(t)$ (dashed line); b) estimation error $e_1(t) = x_1(t) - \hat{x}_1(t)$ for $R = 50$ bit/s, $(T_{s,1} = 0.03$ s, $T_{s,2} = 0.06$ s.

7.6   Conclusions

We have studied dependence of the error of state estimation for nonlinear Lurie systems over a limited-band communication channel both analytically and numerically. It is shown that in common with SISO systems, upper and lower bounds for limit estimation error depend linearly on the transmission error which, in turn, is proportional to the driving signal rate and inversely proportional to the transmission rate. Though these results are obtained for a special type of coder, it reflects peculiarity of the estimation problem as a nonequilibrium dynamical problem. On the contrary, the stabilisation problem considered previously in the literature on control under information constraints belongs to a class of equilibrium problems.

Acknowledgments

The work was supported by NICTA Victoria Research Laboratory of the University of Melbourne, the Russian Foundation for Basic Research (projects RFBR 08-01-00775, 09-08-00803), the Council for grants of the RF President to support young Russian researchers and leading scientific schools (project NSh-2387.2008.1), and the Program of basic research of

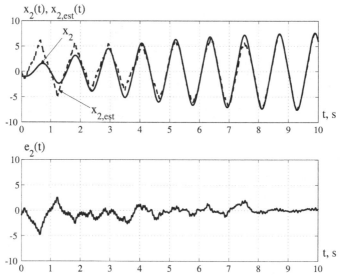

Fig. 7.3   Time histories: a) $x_2(t)$ (solid line), $\hat{x}_2(t)$ (dashed line); b) estimation error $e_2(t) = x_2(t) - \hat{x}_2(t)$ for $R = 50$ bit/s, $(T_{s,1} = 0.03$ s, $T_{s,2} = 0.06$ s.

OEMPPU RAS #2 "Control and safety in energy and technical systems".

References

Abdallah, C. T. and Tanner, H. G. (2007). Complex networked control systems: Introduction to the special section, *Control Systems Magazine, IEEE* **27**, 4, pp. 3–32.

Bazzi, L. M. J. and Mitter, S. K. (2005). Endcoding complexity versus minimum distance, *IEEE Trans. Inform. Theory* **51**, 6, pp. 2103–2112.

Brockett, R. W. and Liberzon, D. (2000). Quantized feedback stabilization of linear systems, *IEEE Trans. Automat. Contr.* **45**, 7, pp. 1279–1289.

Cheng, T. M. and Savkin, A. V. (2007). Output feedback stabilisation of nonlinear networked control systems with non-decreasing nonlinearities: A matrix inequalities approach, *Int. J. Robust Nonlinear Control* **17**, pp. 387–404.

De Persis, C. (2005). n-Bit stabilization of n-dimensional nonlinear systems in feedforward form, *IEEE Trans. Automat. Contr.* **50**, 3, pp. 299–311.

De Persis, C. (2006). On stabilization of nonlinear systems under data rate constraints using output measurements, *Int. J. Robust Nonlinear Control* **16**, pp. 315–332.

De Persis, C. and Isidori, A. (2004). Stabilizability by state feedback implies stabilizability by encoded state feedback, *Systems & Control Letters* **53**, pp. 249–258.

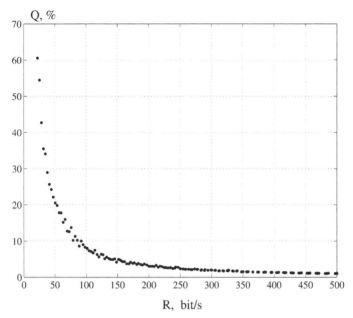

Fig. 7.4 Estimation error Q vs transmission rate $\bar{R}$.

Fradkov, A. L., Andrievsky, B. and Evans, R. J. (2006). Chaotic observer-based synchronization under information constraints, *Physical Review E* **73**, p. 066209.

Goodwin, G. C., Haimovich, H., Quevedo, D. E. and Welsh, J. S. (2004). A moving horizon approach to networked control system design, *IEEE Trans. Automat. Contr.* **49**, 9, pp. 1427–1445.

Ishii, H. and Francis, B. A. (2002). *Limited Data Rate in Control Systems With Networks* (Springer-Verlag, New York).

Liberzon, D. and Hespanha, J. P. (2005). Stabilization of nonlinear systems with limited information feedback, *IEEE Trans. Automat. Contr.* **50**, 6, pp. 910–915.

Matveev, A. S. and Savkin, A. V. (2008). *Estimation and Control over Communication Networks* (Birkhäuser, Boston).

Morgül, O. and Solak, E. (1996). Observer based synchronization of chaotic systems, *Phys. Rev. E* **54**, 5, pp. 4803–4811.

Nair, G. N. and Evans, R. J. (2003). Exponential stabilisability of finite-dimensional linear systems with limited data rates, *Automatica* **39**, pp. 585–593.

Nair, G. N., Evans, R. J., Mareels, I. and Moran, W. (2004). Topological feedback entropy and nonlinear stabilization, *IEEE Trans. Automat. Contr.* **49**, 9, pp. 1585–1597.

Nair, G. N., Fagnani, F., Zampieri, S. and Evans, R. (2007). Feedback control under data rate constraints: an overview, *Proc. IEEE* **95**, 1, pp. 108–137.

Nijmeijer, H. (2001). A dynamical control view on synchronization, *Physica D* **154**, 3–4, pp. 219–228.

Nijmeijer, H. and Mareels, I. M. Y. (1997). An observer looks at synchronization, *IEEE Trans. on Circuits and Systems I* **44**, 10, pp. 882–890.

Savkin, A. V. (2006). Analysis and synthesis of networked control systems: Topological entropy, observability, robustness and optimal control, *Automatica* **42**, pp. 51–62.

Shim, H., Seo, J. H. and Teel, A. R. (2003). Nonlinear observer design via passification of error dynamics, *Automatica* **39**, pp. 885–892.

Tatikonda, S. and Mitter, S. (2004). Control under communication constraints, *IEEE Trans. Automat. Contr.* **49**, 7, pp. 1056–1068.

Wong, W. S. and Brockett, R. W. (1997). Systems with finite communication bandwidth constraints – Part I: State estimation problems, *IEEE Trans. Automat. Contr.* **42**, 9, pp. 1294–1299.

Tracking Control of Multiconstraint Nonsmooth Lagrangian Systems

C. Morarescu, B. Brogliato, T. Nguyen

INRIA, Grenoble Rhone-Alpes
France

Abstract

In this study one considers the tracking control problem of a class of nonsmooth fully actuated Lagrangian systems subject to frictionless unilateral constraints. The task under consideration consists of a succession of free and constrained phases. The transition from a constrained to a free phase is monitored via a Linear Complementarity Problem (LCP). On the other hand during the transition from a free to a constrained phase the dynamics contains some impacts that hamper the asymptotic stability. Nevertheless we have proved the practical weak stability of the system with an almost decreasing Lyapunov function. One numerical example illustrates the methodology described in the paper.

8.1 Introduction

The control of mechanical systems subject to unilateral constraints has been the object of many studies in the past fifteen years. Such systems, which consist of three main ingredients (see (8.1) below) are highly nonlinear nonsmooth dynamical systems. Theoretical aspects of their Lyapunov stability and the related stabilization issues have been studied in [Brogliato, 2004; Leine and van de Wouw, 2008c; Leine and van de Wouw, 2008a; Tornambé,

1999]. The specific yet important task of the stabilization of impacting transition phases was analyzed and experimentally tested in [Lee et al., 2003; Pagilla and Yu, 2004; Sekhavat et al., 2006; van Vliet et al., 2000; Volpe and and Khosla, 1993; Xu et al., 2000]. From the point of view of tracking control of complementarity Lagrangian systems along general constrained/unconstrained paths, such studies focus on a module of the overall control problem. One of the first works formulating the control of complete robotic tasks via unilateral constraints and complementarity conditions was presented in [Huang and McClamroch, 1988]. In that work the impacts were considered inelastic and the control problem was solved using a time optimal problem. The tracking control problem under consideration, involving systems that undergo transitions from free to constrained motions, and vice-versa, along an infinity of cycles, was formulated and studied in [Brogliato et al., 1997] mainly for the 1-dof (degree-of-freedom) case and in [Bourgeot and Brogliato, 2005] for the n-dof case. Both of these works consider systems with only one unilateral frictionless constraint. In this chapter we not only consider the multiconstraint case but the results in Section 8.7 relax some very hard to verify conditions imposed in [Bourgeot and Brogliato, 2005] to assure the stability. Moreover the accurate design of the control law that guarantees the detachment from the constraints is formulated and incorporated in the stability analysis for the first time. Considering multiple constraints may be quite important in applications like virtual reality and haptic systems, where typical tasks involve manipulating objects modelled as rigid bodies [Faulring et al., 2007] in complex environments with many unilateral constraints. We note that in the case of a single nonsmooth impact the exponential stability and bounded-input bounded state (BIBS) stability was studied in [Menini and Tornambè, 2001b] using a state feedback control law. A study for a multiple degree-of-freedom linear systems subject to nonsmooth impacts can be found in [Menini and Tornambè, 2001c]. That approach proposes a proportional-derivative control law in order to study BIBS stability via Lyapunov techniques. Other approaches for the tracking control of nonsmooth mechanical systems can be found in [Galeani et al., 2008; Menini and Tornambè, 2001a; Pagilla, 2001] and in [Leine and van de Wouw, 2008b]. The analysis and control of systems subject to unilateral constraints also received attention in [Bentsman and Miller, 2007].

This chapter focuses on the problem of tracking control of complementarity Lagrangian systems [Moreau, 1988] subject to frictionless unilateral constraints whose dynamics may be expressed as:

$$\begin{cases} M(X)\ddot{X} + C(X,\dot{X})\dot{X} + G(X) = U + \nabla F(X)\lambda_X \\ 0 \leq \lambda_X \perp F(X) \geq 0, \\ \text{Collision rule} \end{cases} \qquad (8.1)$$

where $X(t) \in \mathbb{R}^n$ is the vector of generalized coordinates, $M(X) = M^T(X) \in \mathbb{R}^{n \times n}$ is the positive definite inertia matrix, $F(X) \in \mathbb{R}^m$ represents the distance to the constraints, $C(X,\dot{X})$ is the matrix containing Coriolis and centripetal forces, $G(X)$ contains conservative forces, $\lambda_X \in \mathbb{R}^m$ is the vector of the Lagrangian multipliers associated to the constraints and $U \in \mathbb{R}^n$ is the vector of generalized torque inputs. For the sake of completeness we precise that ∇ denotes the Euclidean gradient $\nabla F(X) = (\nabla F_1(X), \ldots, \nabla F_m(X)) \in \mathbb{R}^{n \times m}$ where $\nabla F_i(X) \in \mathbb{R}^n$ represents the vector of partial derivatives of $F_i(\cdot)$ w.r.t. the components of X. We assume that the functions $F_i(\cdot)$ are continuously differentiable and that $\nabla F_i(X) \neq 0$ for all X with $F_i(X) = 0$. It is worth to precise here that for a given function $f(\cdot)$ its derivative w.r.t. the time t will be denoted by $\dot{f}(\cdot)$. For any function $f(\cdot)$ the limit to the right at the instant t will be denoted by $f(t^+)$ and the limit to the left will be denoted by $f(t^-)$. A simple jump of the function $f(\cdot)$ at the moment $t = t_\ell$ is denoted $\sigma_f(t_\ell) = f(t_\ell^+) - f(t_\ell^-)$. The Dirac measure at time t is δ_t.

Definition 8.1. A Linear Complementarity Problem (LCP) is a system given by:

$$0 \leq \lambda \perp A\lambda + b \geq 0 \qquad (8.2)$$

Such an LCP has a unique solution for all b if and only if A is a P-matrix, i.e. all its principal minors are positive [Facchinei and Pang, 2003].

The admissible domain associated to the system (8.1) is the closed set Φ where the system can evolve and it is described as follows:

$$\Phi = \{X \in \mathbb{R}^n \mid F(X) \geq 0\} = \bigcap_{1 \leq i \leq m} \Phi_i,$$

where $\Phi_i = \{X \in \mathbb{R}^n \mid F_i(X) \geq 0\}$ considering that a vector is non-negative if and only if all its components are non-negative. In order to have a well-posed problem with a physical meaning we consider that Φ contains at least a closed ball of positive radius.

Definition 8.2. A singularity of the boundary $\partial\Phi$ of Φ is the intersection of two or more codimension one surfaces $\Sigma_i = \{X \in \mathbb{R}^n \mid F_i(X) = 0\}$.

The presence of $\partial\Phi$ may induce some impacts that must be included in the dynamics of the system. It is obvious that $m > 1$ allows both simple impacts (when one constraint is involved) and multiple impacts (when singularities or surfaces of codimension larger than 1 are involved). Let us introduce the following notion of p_ϵ-impact.

Definition 8.3. Let $\epsilon \geq 0$ be a fixed real number. We say that a p_ϵ-impact occurs at the instant t if

$$||F_I(X(t))|| \leq \epsilon, \quad \prod_{i \in I} F_i(X(t)) = 0$$

where $I \subset 1, \ldots, m$, $card(I) = p$, $F_I(X) = (F_{i_1}(X), F_{i_2}(X), ..., F_{i_p}(X))^T$, $i_1, i_2, ..., i_p \in I$.

If $\epsilon = 0$ the p surfaces Σ_i, $i \in I$ are stroked simultaneously and a p-impact occurs. When $\epsilon > 0$ the system collides $\partial\Phi$ in a neighborhood of the intersection $\bigcap_{i \in I} \Sigma_i$ (see Figure 8.1).

Fig. 8.1   Illustration of 2_ϵ-impacts.

Definition 8.4 (Moreau, 1988; Hiriart-Urruty & Lemaréchal, 2001). *The tangent cone to* $\Phi = \{X \in \mathbb{R}^n \mid F_i(X) \geq 0, \forall i = 1, \ldots, m\}$ *at* $X \in \mathbb{R}^n$ *is defined as:*

$$T_\Phi(X) = \{z \in \mathbb{R}^n \mid z^T \nabla F_i(X) \geq 0, \forall i = J(X)\}$$

where $J(X) \triangleq \{i \in \{1, \ldots, m\} \mid F_i(X) \leq 0\}$ *is the index set of active constraints. When* $X \in \Phi \setminus \partial\Phi$ *one has* $J(X) = \emptyset$ *and* $T_\Phi(X) = \mathbb{R}^n$. *The normal cone to* Φ *at* X *is defined as the polar cone to* $T_\Phi(X)$:

$$N_\Phi(X) = \{y \in \mathbb{R}^n \mid \forall z \in T_\Phi(X), y^T z \leq 0\}$$

The collision (or restitution) rule in (8.1), is a relation between the post-impact velocity and the pre-impact velocity. Among the various models of collision rules, Moreau's rule is an extension of Newton's law which is energetically consistent [Glocker, 2004; Mabrouk, 1998] and is numerically tractable [Acary and Brogliato, 2008]. For these reasons throughout this chapter the collision rule will be defined by Moreau's relation [Moreau, 1988]:

$$\dot{X}(t_\ell^+) = (1+e) \mathop{\arg\min}_{z \in T_\Phi(X(t_\ell))} \frac{1}{2}[z - \dot{X}(t_\ell^-)]^T M(X(t_\ell))[z - \dot{X}(t_\ell^-)] - e\dot{X}(t_\ell^-)$$

(8.3)

where $\dot{X}(t_\ell^+)$ is the post-impact velocity, $\dot{X}(t_\ell^-)$ is the pre-impact velocity and $e \in [0,1]$ is the restitution coefficient. Denoting by T the kinetic energy of the system, we can compute the kinetic energy loss at the impact time t_ℓ as [Mabrouk, 1998]:

$$T_L(t_\ell) = -\frac{1-e}{2(1+e)}\left[[\dot{X}(t_\ell^+) - \dot{X}(t_\ell^-)]^T M(X(t_\ell))[\dot{X}(t_\ell^+) - \dot{X}(t_\ell^-)]\right] \leq 0$$

(8.4)

The collision rule can be rewritten considering the vector of generalized velocities as an element of the tangent space to the configuration space of the system, equipped with the kinetic energy metric. Doing so (see [Brogliato, 1999] §6.2), the discontinuous velocity components $\dot{X}_{norm}$ and the continuous ones $\dot{X}_{tang}$ are identified. Precisely, $\begin{pmatrix} \dot{X}_{norm} \\ \dot{X}_{tang} \end{pmatrix} =$

$\mathcal{M}\dot{X}$, $\mathcal{M} = \begin{pmatrix} \mathbf{n} \\ \mathbf{t} \end{pmatrix} M(X)$ where $\mathbf{n}$ represents the unitary normal vec-

tors $\mathbf{n}_i = \dfrac{M^{-1}(X(t_\ell))\nabla F_i(X(t_\ell))}{\sqrt{\nabla F_i(X(t_\ell))^T M^{-1}(X(t_\ell))\nabla F_i(X(t_\ell))}}$, $i \in J(X(t_\ell))$ (see Definition 8.4) and $\mathbf{t}$ represents mutually independent unitary vectors $\mathbf{t}_i$ such that $\mathbf{t}_i^T M(X(t_\ell))\mathbf{n}_j = 0, \forall i,j$. In this case the collision rule (8.3) at the impact time t_ℓ becomes the generalized Newton's rule $\begin{pmatrix} \dot{X}_{norm}(t_\ell^+) \\ \dot{X}_{tang}(t_\ell^+) \end{pmatrix} =$

$-\eta \begin{pmatrix} \dot{X}_{norm}(t_\ell^-) \\ \dot{X}_{tang}(t_\ell^-) \end{pmatrix}$, $\eta = diag(e_1, \ldots, e_m, 0, \ldots, 0)$ where e_i is the restitu-

tion coefficient w.r.t. the surface Σ_i. For the sake of simplicity we consider in this chapter that all the restitution coefficients are equal, i.e. $e_1 = \ldots = e_m \triangleq e$.

The structure of the chapter is as follows: in Section 8.2 one presents some basic concepts and prerequisites necessary for the further developments. Section 8.3 is devoted to the controller design. In Section 8.4 one defines the exogenous trajectories entering the dynamics. The desired contact-force that must occur on the phases where the motion is constrained, is explicitly defined in Section 8.5. Section 8.6 focuses on the strategy for take-off at the end of the constraint phases. The main results related to the closed-loop stability analysis are presented in Section 8.7. Examples and concluding remarks end the chapter.

The following standard notations will be adopted: $||\cdot||$ is the Euclidean norm, $b_p \in \mathbb{R}^p$ and $b_{n-p} \in \mathbb{R}^{n-p}$ are the vectors formed with the first p and the last $n - p$ components of $b \in \mathbb{R}^n$, respectively. $N_\Phi(X_p = 0)$ is the normal cone $N_\Phi(X)$ to Φ at X when X satisfies $X_p = 0$, $\lambda_{min}(\cdot)$ and $\lambda_{max}(\cdot)$ represent the smallest and the largest eigenvalues, respectively.

8.2   Basic concepts

8.2.1   Typical task

Following [Brogliato et al., 1997] the time axis can be split into intervals Ω_k and I_k corresponding to specific phases of motion. Due to the singularities of $\partial\Phi$ that must be taken into account, the constrained-motion phases need to be decomposed in sub-phases where some specific constraints are active. Between two such sub-phases a transition phase occurs only when the number of active constraints increases. This means that a typical task can be represented in the time domain as:

$$t \in \mathbb{R}^+ = \Omega_0^\emptyset \cup \left[\bigcup_{k \geq 1} \left(I_k \cup \left(\bigcup_{i=1}^{m_k} \Omega_k^{J_{k,i}} \right) \right) \right] \tag{8.5}$$

$$J_{k,m_k} \subset J_{k+1,1}, J_{k,m_k} \subset J_{k,m_k-1} \subset \ldots \subset J_{k,1}$$

where the superscript $J_{k,i} = \{j \in \{1,\ldots,m\} \mid F_j(X) = 0\}$ represents the set of active constraints during the corresponding motion phase, and I_k denotes the transient between two Ω_k phases when the number of active constraints increases. Without loss of generality we suppose that the system is initialized in the interior of Φ at a free-motion phase. The impacts during I_k involve $p = |J_{k,1}|$ constraints (p_ϵ-impacts). Furthermore we shall prove that the first impact of I_k is a p_ϵ-impact with ϵ bounded by a parameter chosen by the designer. When the number of active constraints decreases there is no impact, thus no other transition phases are needed. We note that $J_{k,i} = \emptyset$ corresponds to free-motion ($F(X) > 0$).

Since the tracking control problem involves no difficulty during the Ω_k phases, *the central issue is the study of the passages between them (the de-*

sign of transition phases I_k and detachment conditions), and the stability of the trajectories evolving along (8.5) (i.e. an infinity of cycles). It is noteworthy that the passage $\Omega_k^{J_{k,i}} \to \Omega_k^{J_{k,i+1}}$ consists of detachments from some constraints. In Section 8.6 we consider that p constraints are active and we give the conditions to smoothly take-off from r of them. It is clear that once we know how to do that, we can manage all the transitions mentioned above. Throughout the chapter, the sequence $I_k \cup \left(\bigcup_{i=1}^{m_k} \Omega_k^{J_{k,i}} \right)$ will be referred to as the cycle k of the system's evolution. For robustness reasons during transition phases I_k we impose a closed-loop dynamics (containing impacts) that mimics somehow the bouncing-ball dynamics (see e.g. [Brogliato, 1999]).

8.2.2 *Exogenous signals entering the dynamics*

In this section we introduce the trajectories playing a role in the dynamics and the design of the controller. Some instants that will be used further are also defined.

- $X^{nc}(\cdot)$ denotes the desired trajectory of the unconstrained system (i.e. the trajectory that the system should track if there were no constraints). We suppose that $F(X^{nc}(t)) < 0$ for some t, otherwise the problem reduces to the tracking control of a system with no constraints.
- $X_d^*(\cdot)$ denotes the signal entering the control input and playing the role of the desired trajectory during some parts of the motion.
- $X_d(\cdot)$ represents the signal entering the Lyapunov function. This signal is set on the boundary $\partial \Phi$ after the first impact of each cycle.

The signals $X_d^*(\cdot)$ and $X_d(\cdot)$ coincide on the Ω_k phases while $X^{nc}(\cdot)$ is used to define everywhere $X_d^*(\cdot)$ and $X_d(\cdot)$.
Throughout the chapter we consider $I_k = [\tau_0^k, t_f^k]$, where τ_0^k is chosen by the designer as the start of the transition phase I_k and t_f^k is the end of I_k. We note that all superscripts $(\cdot)^k$ will refer to the cycle k of the system motion. We also use the following notations:

- t_0^k is the first impact during the cycle k,
- t_∞^k is the accumulation point of the sequence $\{t_\ell^k\}_{\ell \geq 0}$ of the impact instants during the cycle k $(t_f^k \geq t_\infty^k)$,
- τ_1^k will be explicitly defined later and represents the instant when the signal $X_d^*(\cdot)$ reaches a given value chosen by the designer in order to impose a closed-loop dynamics with impacts during the transition phases,

- $t_d^{k,i}$ is the desired detachment instant at the end of the phase $\Omega_k^{J_{k,i}}$.

It is noteworthy that t_0^k, t_∞^k are state-dependent whereas τ_0^k, τ_1^k and $t_d^{k,i}$ are exogenous and imposed by the designer. To better understand the definition of these specific instants, in Figure 8.2 we represent the exogenous signals $X^{nc}(\cdot), X_d(\cdot), X_d^*(\cdot)$ during a sequence $\Omega_{k-1}^{J_{k-1}} \cup I_k \cup \Omega_k^{J_{k,1}} \cup \Omega_k^{J_{k,2}}$ when the motion is simplified as follows:

- during the transition phase we take into account only the constraints that must be activated $J_{k,1} \setminus J_{k-1,m_{k-1}}$.
- at the end of the phase $\Omega_k^{J_{k,1}}$ we take into account only the constraints that must be deactivated $J_{k,1} \setminus J_{k,2}$.

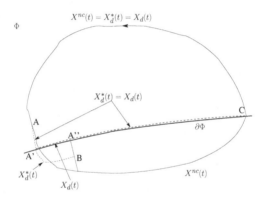

Fig. 8.2   The closed-loop desired trajectory and control signals.

The points A, A', A'' and C in Figure 8.2 correspond to the moments τ_0^k, t_0^k, t_f^k and $t_d^{k,1}$ respectively. We have seen that the choice of τ_0^k plays an important role in the stability criterion given by Proposition 8.1. On the other hand in Figure 8.2 we see that starting from A the desired trajectory $X_d(\cdot) = X_d^*(\cdot)$ is deformed compared to $X^{nc}(\cdot)$. In order to reduce this deformation, the time τ_0^k and implicitly the point A must be close to $\partial\Phi$. Further details on the choice of τ_0^k will be given later. Taking into account just the constraints $J_{k,1} \setminus J_{k,2}$ we can identify $t_d^{k,1}$ with the moment when $X_d(\cdot)$ and $X^{nc}(\cdot)$ rejoin at C.

In order to clarify the differences and the usefulness of the signals introduced above we present a simple example concerning a one-degree of freedom model.

Example 8.1. Let us consider the following one degree-of-freedom dynamical system:

$$\begin{cases} (\ddot{X} - \ddot{X}_d^*) + \gamma_2(\dot{X} - \dot{X}_d^*) + \gamma_1(X - X_d^*) = \lambda \\ 0 \leq X \perp \lambda \geq 0 \\ \dot{X}(t_k^+) = -e_n\dot{X}(t_k^-) \end{cases} \tag{8.6}$$

where $X_d^*(\cdot)$ is a twice differentiable signal and γ_1, γ_2 are two gains. The real number λ represents the Lagrange multiplier associated to the constraint $X \geq 0$.

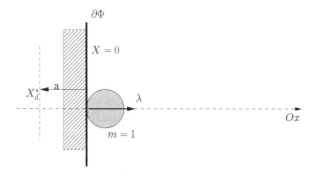

Fig. 8.3   One degree of freedom: unitary mass dynamics.

The system (8.6) represents the dynamics of a unitary mass lying in the right half plane ($X \geq 0$) and moving on the Ox axis (see figure 8.3). Thus, the system dynamics is expressed by $\ddot{X} = U + \lambda$ with the control law $U = \ddot{X}_d^* - \gamma_2(\dot{X} - \dot{X}_d^*) - \gamma_1(X - X_d^*)$. One considers that the Lyapunov function $V(X, \dot{X}, t)$ of the closed-loop system is given by a quadratic function of the tracking error. In order to stabilize the system on the constraint surface $\partial\Phi$ there are two solutions.

[**1st solution**] Setting $X_d^*(t) = 0$, $\dot{X}_d^*(t) = 0$ and $\ddot{X}_d^*(t) = 0$ at the time t of contact with $\partial\Phi$. Therefore when the system reaches its equilibrium point on $\partial\Phi$ one obtains the contact force $\lambda = 0$. It is noteworthy that, since we are interested to study a dynamics containing a constraint movement phase, we need a non-zero contact force when the system is stabilized on $\partial\Phi$. Therefore this type of solution is not convenient for the study developed in this work. Moreover the stabilization on $\partial\Phi$ with a non zero tracking error on $[t - \epsilon, t]$ is not obvious, especially in higher dimensions, so that the conditions for a perfect tangential approach are never met in practice.

[**2nd solution**] Setting $X_d^*(t) = -a$, $a > 0$, $\dot{X}_d^*(t) = 0$ and $\ddot{X}_d^*(t) = 0$ on some interval of time when approaching $\partial\Phi$. Since the tracking error $\bar{X} =$

$X - X_d^* = -X_d^*$ at the equilibrium point we cannot use $\bar{X}$ in the definition of the Lyapunov function. Obviously the system reaches the desired position when $X = 0$. Thus we shall use the signal $X_d(t) = 0$ in order to express the desired trajectory in the Lyapunov function and the corresponding tracking error will be given by $\tilde{X} = X - X_d$. Concluding, when the equilibrium point is reached one obtains $\tilde{X} = 0$, $\bar{X} = -a$, $V(t, \tilde{X} = 0) = 0$ and the corresponding contact force $\lambda = -\gamma_1 X_d^* = \gamma_1 a > 0$.

8.2.3 Stability analysis criteria

The system (8.1) is a complex nonsmooth and nonlinear dynamical system which involves continuous and discrete time phases. A stability framework for this type of systems has been proposed in [Brogliato et al., 1997] and extended in [Bourgeot and Brogliato, 2005]. This is an extension of the Lyapunov second method adapted to closed-loop mechanical systems with unilateral constraints. Since we use this criterion in the following tracking control strategy it is worth to clarify the framework and to introduce some definitions.

Let us define Ω as the complement in $\mathbb{R}^+$ of $I = \bigcup_{k \geq 1} I_k$ and assume that the Lebesgue measure of Ω, denoted $\lambda[\Omega]$, equals infinity. Consider $x(\cdot)$ the state of the closed-loop system in (8.1) with some feedback controller $U(X, \dot{X}, X_d^*, \dot{X}_d^*, \ddot{X}_d^*)$.

Definition 8.5 (Weakly Stable System). *The closed loop system is called weakly stable if for each $\epsilon > 0$ there exists $\delta(\epsilon) > 0$ such that $||x(0)|| \leq \delta(\epsilon) \Rightarrow ||x(t)|| \leq \epsilon$ for all $t \geq 0$, $t \in \Omega$. The system is asymptotically weakly stable if it is weakly stable and $\lim_{t \in \Omega, t \to \infty} x(t) = 0$. Finally, the practical weak stability holds if there exists $0 < R < +\infty$ and $t^* < +\infty$ such that $||x(t)|| < R$ for all $t > t^*$, $t \in \Omega$.*

Weak stability is therefore Lyapunov stability without looking at the transition phases. Consider $V(\cdot)$ such that there exists strictly increasing functions $\alpha(\cdot)$ and $\beta(\cdot)$ satisfying the conditions: $\alpha(0) = 0$, $\beta(0) = 0$ and $\alpha(||x||) \leq V(x, t) \leq \beta(||x||)$.

Definition 8.6. *A transition phase I_k is called finite if it involves a sequence of impact times $(t_\ell^k)_{0 \leq \ell \leq N}$, $N \leq \infty$ with the accumulation point $t_N^k < \infty$ (for the sake of simplicity we shall denote the accumulation point by t_∞^k even if $N < \infty$).*

In the sequel all the transition phases are supposed finite, which implies that $e < 1$ (in [Ballard, 2001] it is shown that $e = 1$ implies that $t_\infty^k = +\infty$). The following criterion is inspired from [Bourgeot and Brogliato, 2005], and will be used to study the stability of the system (8.1).

Proposition 8.1 (Weak Stability). *Assume that the task admits the representation (8.5) and that*

(a) $\lambda[I_k] < +\infty, \quad \forall k \in \mathbb{N}$,
(b) outside the impact accumulation phases $[t_0^k, t_\infty^k]$ one has $\dot{V}(x(t), t) \leq -\gamma V(x(t), t)$ for some constant $\gamma > 0$,
(c) $\sum_{\ell \geq 0} \left[V(t_{\ell+1}^{k-}) - V(t_\ell^{k+}) \right] \leq K_1 V^{p_1}(\tau_0^k), \forall k \in \mathbb{N}$ for some $p_1 \geq 0$, $K_1 \geq 0$,
(d) the system is initialized on Ω_0 such that $V(\tau_0^1) \leq 1$,
(e) $\sum_{\ell \geq 0} \sigma_V(t_\ell^k) \leq K_2 V^{p_2}(\tau_0^k) + \xi, \forall k \in \mathbb{N}$ for some $p_2 \geq 0$, $K_2 \geq 0$ and $\xi \geq 0$.

If $p = \min\{p_1, p_2\} < 1$ then $V(\tau_0^k) \leq \delta(\gamma, \xi), \forall k \geq 2$, where $\delta(\gamma, \xi)$ is a function that can be made arbitrarily small by increasing the value of γ. The system is practically weakly stable with $R = \alpha^{-1}(\delta(\gamma, \xi))$.

Remark 8.1. Since the Lyapunov function is exponentially decreasing on the Ω_k phases, assumption **(d)** in Proposition 8.1 means that the system is initialized on Ω_0 sufficiently far from the moment when the trajectory $X^{nc}(\cdot)$ leaves the admissible domain.

Precisely, the weak stability is characterized by an "almost decreasing" Lyapunov function $V(x(\cdot).\cdot)$ as illustrated in Figure 8.4.

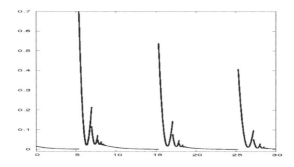

Fig. 8.4   Typical evolution of the Lyapunov function of weakly stable systems.

Remark 8.2. It is worth to point out the local character of the stability criterion proposed by Proposition 8.1. This character is firstly given by condition **d)** of the statement and secondly by the synchronization constraints of the control law and the motion phase of the system (see (8.5) and (8.7) below).

The practical stability is very useful because attaining asymptotic stability is not an easy task for the unilaterally constrained systems described by (8.1) especially when $n \geq 2$ and $M(q)$ is not a diagonal matrix (i.e. there are inertial couplings, which is the general case).

8.3   Controller design

In order to overcome some difficulties that can appear in the controller definition, the dynamical equations (8.1) will be expressed in the generalized coordinates introduced by McClamroch and Wang (1988), which allow one to split the generalized coordinates into a "normal" and a "tangential" parts, with a suitable diffeomorphic transformation $q = Q(X)$. We suppose that the generalized coordinates transformation holds globally in Φ, which may obviously not be the case in general. However, the study of the singularities that might be generated by the coordinates transformation is out of the scope of this chapter. Let us consider $D = [I \vdots O] \in \mathbb{R}^{m \times n}$, $I \in \mathbb{R}^{m \times m}$ the identity matrix. The new coordinates will be $q = Q(X) \in \mathbb{R}^n$, with

$$q = \begin{bmatrix} q_1 \\ q_2 \end{bmatrix}, \ q_1 = \begin{bmatrix} q_1^1 \\ \vdots \\ q_1^m \end{bmatrix} \text{ such that } \Phi = \{q \mid Dq \geq 0\}^1.$$ The tangent cone $T_\Phi(q_1 = 0) = \{v \mid Dv \geq 0\}$ is the space of admissible velocities on the boundary of Φ.

The controller used here consists of different low-level control laws for each phase of the system. More precisely, the switching controller can be expressed as

$$T(q)U = \begin{cases} U_{nc} \text{ for } t \in \Omega_k^\emptyset \\ U_t^J \text{ for } t \in I_k \\ U_c^J \text{ for } t \in \Omega_k^J \end{cases} \tag{8.7}$$

---

[1]In particular it is implicitly assumed that the functions $F_i(\cdot)$ in (8.1) are linearly independent.

where $T(q) = \begin{pmatrix} T_1(q) \\ T_2(q) \end{pmatrix} \in \mathbb{R}^{n \times n}$ is full-rank under some basic assumptions (see [McClamroch and Wang, 1988]). The dynamics becomes:

$$\begin{cases} M_{11}(q)\ddot{q}_1 + M_{12}(q)\ddot{q}_2 + C_1(q,\dot{q})\dot{q} + g_1(q) = T_1(q)U + \lambda \\ M_{21}(q)\ddot{q}_1 + M_{22}(q)\ddot{q}_2 + C_2(q,\dot{q})\dot{q} + g_2(q) = T_2(q)U \\ q_1^i \geq 0,\ q_1^i \lambda_i = 0,\ \lambda_i \geq 0,\ 1 \leq i \leq m \\ \text{Collision rule} \end{cases} \quad (8.8)$$

where the set of complementary relations can be written more compactly as $0 \leq \lambda \perp Dq \geq 0$.

In the sequel U_{nc} coincides with the fixed-parameter controller proposed in [Slotine and Li, 1988] and the closed-loop stability analysis of the system is based on Proposition 8.1. First, let us introduce some notations: $\tilde{q} = q - q_d$, $\bar{q} = q - q_d^*$, $s = \dot{\tilde{q}} + \gamma_2 \tilde{q}$, $\bar{s} = \dot{\bar{q}} + \gamma_2 \bar{q}$, $\dot{q}_e = \dot{q}_d - \gamma_2 \tilde{q}$ where $\gamma_2 > 0$ is a scalar gain and $q_d(\cdot)$, $q_d^*(\cdot)$ represent the desired trajectories defined in the previous section. Using the above notations the controller is given by

$$T(q)U \triangleq \begin{cases} U_{nc} = M(q)\ddot{q}_e + C(q,\dot{q})\dot{q}_e + G(q) - \gamma_1 s \\ U_t^J = U_{nc},\ t \leq t_0^k \\ U_t^J = M(q)\ddot{q}_e + C(q,\dot{q})\dot{q}_e + G(q) - \gamma_1 \bar{s},,\ t > t_0^k \\ U_c^J = U_{nc} - P_d + K_f(P_q - P_d) \end{cases} \quad (8.9)$$

where $\gamma_1 > 0$ is a scalar gain, $K_f > 0$, $P_q = D^T \lambda$ and $P_d = D^T \lambda_d$ is the desired contact force during persistently constrained motion. It is clear that during Ω_k^J not all the constraints are active and, therefore, some components of λ and λ_d are zero.

In order to prove the stability of the closed-loop system (8.7)–(8.9) we will use the following positive definite function:

$$V(t, s, \tilde{q}) = \frac{1}{2} s^T M(q) s + \gamma_1 \gamma_2 \tilde{q}^T \tilde{q} \quad (8.10)$$

8.4   Tracking control framework

8.4.1   *Design of the desired trajectories*

In this chapter we treat the tracking control problem for the closed-loop dynamical system (8.7)–(8.9) with the complete desired path a priori taking into account the complementarity conditions and the impacts. In order to define the desired trajectory let us consider the motion of a virtual and unconstrained particle perfectly following a trajectory (represented by $X^{nc}(\cdot)$ on Figure 8.2) with an orbit that leaves the admissible domain for a given period. Therefore, the orbit of the virtual particle can be split into two parts, one of them belonging to the admissible domain (inner part) and the other one outside the admissible domain (outer part). In the sequel we deal with the tracking control strategy when the desired trajectory is constructed such that:

(i) when no activated constraints, it coincides with the trajectory of the virtual particle (the desired path and velocity are defined by the path and velocity of the virtual particle, respectively),

(ii) when $p \leq m$ constraints are active, its orbit coincides with the projection of the outer part of the virtual particle's orbit on the surface of codimension p defined by the activated constraints (X_d between A'' and C in Figure 8.2),

(iii) the desired detachment moment and the moment when the virtual particle re-enters the admissible domain (with respect to $p \leq m$ constraints) are synchronized.

Therefore we have not only to track a desired path but also to impose a desired velocity allowing the motion synchronization on the admissible domain. The main difficulties here consist of:

- stabilizing the system on $\partial\Phi$ during the transition phases I_k and incorporating the velocity jumps in the overall stability analysis;
- deactivating some constraints at the moment when the unconstrained trajectory re-enters the admissible domain with respect to them;
- maintaining a persistently constrained motion between the moment when the system was stabilized on $\partial\Phi$ and the detachment moment.

Remark 8.3. The problem can be relaxed considering that we want to track only a desired path like $X^{nc}(\cdot)$ (without imposing a desired velocity on the inner part of the desired trajectory and/or a given period to complete a cycle). In this way the synchronization problem (iii) disappears and we can assume there exists a twice differentiable desired trajectory outside $[t_0^k, t_f^k]$ that assures the detachment when the force control is dropped. In other words, in this case we have to design the desired trajectory only during I_k phases.

8.4.2 Design of $q_d^*(\cdot)$ and $q_d(\cdot)$ on the phases I_k

During the transition phases the system must be stabilized on $\partial\Phi$. Obviously, this does not mean that all the constraints have to be activated (i.e.

$q_1^i(t) = 0, \forall i = 1, \dots, m)$. Let us consider that only the first p constraints (eventually reordering the coordinates) define the border of Φ where the system must be stabilized. The following methodology will be used to define $q_d^*(\cdot)$:

1) During a small period $\delta > 0$ chosen by the designer the desired velocity becomes zero preserving the twice differentiability of $q_d^*(\cdot)$. For instance we can use the following definition:

$$q_d^*(t) = q^{nc}\left(\tau_0^k + \frac{(t - \tau_0^k - \delta)^2(t - \tau_0^k)}{\delta^2}\right), t \in [\tau_0^k, \tau_0^k + \delta]$$

which means $q_d^*(\tau_0^k + \delta) = q_d^*(\tau_0^k) = q^{nc}(\tau_0^k)$, $\dot{q}_d^*(\tau_0^k + \delta) = 0$ and $\dot{q}_d^*(\tau_0^k) = \dot{q}^{nc}(\tau_0^k)$

2) The last $n - p$ components of $q_d^*(\cdot)$ are frozen:

$$(q_d^*)_{n-p}(t) = q_{n-p}^{nc}(\tau_0^k), \quad t \in (\tau_0^k + \delta, t_f^k] \tag{8.11}$$

3) For a fixed $\varphi > 0$ the moment τ_1^k is chosen by the designer as the instant when the limit conditions $(q_d^i)^*(\tau_1^k) = -\nu V^{1/3}(\tau_0^k)$, $(\dot{q}_d^i)^*(\tau_1^k) = 0, \forall i = 1, \dots, p$, hold. On $[\tau_0^k + \delta, \tau_1^k)$ we define $q_d^*(\cdot)$ as a twice differentiable decreasing signal. Precisely, denoting $t' = \frac{t - (\tau_0^k + \delta)}{\tau_1^k - (\tau_0^k + \delta)}$, the components $(q_d^i)^*(\cdot)$, $i = 1, \dots, p$ of $(q_d^*)_p(\cdot)$ are defined as:

$$(q_d^i)^*(t) = \begin{cases} a_3^i(t')^3 + a_2^i(t')^2 + a_0^i, & t \in [\tau_0^k + \delta, \min\{\tau_1^k; t_0^k\}] \\ -\varphi V^{1/3}(\tau_0^k), & t \in (\min\{\tau_1^k; t_0^k\}, t_f^k] \end{cases} \tag{8.12}$$

where $V(\cdot)$ is defined in (8.10) and the coefficients are:

$$\begin{align} a_3^i &= 2[(q^i)^{nc}(\tau_0^k) + \varphi V^{1/3}(\tau_0^k)] \\ a_2^i &= -3[(q^i)^{nc}(\tau_0^k) + \varphi V^{1/3}(\tau_0^k)] \tag{8.13} \\ a_0^i &= (q^i)^{nc}(\tau_0^k) \end{align}$$

The rationale behind the choice of $q_d^*(\cdot)$ is on one hand to assure a robust stabilization on $\partial\Phi$, mimicking the bouncing-ball dynamics; on the other hand to enable one to compute suitable upper-bounds that will help using Proposition 8.1 (hence $V^{1/3}(\cdot)$ terms in (8.12) with $V(\cdot)$ in (8.10)).

Remark 8.4. Two different situations are possible. The first one is given by $t_0^k > \tau_1^k$ and we shall prove that in this situation all the jumps of the Lyapunov function in (8.10) are negative. The second situation was pointed out in [Bourgeot and Brogliato, 2005] and is given by $t_0^k < \tau_1^k$. In this situation the first jump at t_0^k in the Lyapunov function may be positive. It is noteworthy that $q_d^*(\cdot)$ will then have a jump at the time t_0^k since $(q_d^i)^*(t_0^{k+}) = -\varphi V^{1/3}(\tau_0^k), \forall i = 1, \dots, p$ (see (8.12)).

In order to limit the deformation of the desired trajectory $q_d^*(\cdot)$ w.r.t. the unconstrained trajectory $q^{nc}(\cdot)$ during the I_k phases (see Figures 8.2), we impose in the sequel

$$||q_p^{nc}(\tau_0^k)|| \leq \psi \tag{8.14}$$

where $\psi > 0$ is chosen by the designer. It is obvious that a smaller ψ leads to smaller deformation of the desired trajectory and to smaller deformation of the real trajectory as we shall see in Section 8.8. Nevertheless, due to the tracking error, ψ cannot be chosen zero. We also note that $||q_p^{nc}(\tau_0^k)|| \leq \psi$ is a practical way to choose τ_0^k.

During the transition phases I_k we define $(q_d)_{n-p}(t) = (q_d^*)_{n-p}(t)$. Assuming a finite accumulation period, the impact process can be considered in some way equivalent to a plastic impact. Therefore, $(q_d)_p(\cdot)$ and $(\dot{q}_d)_p(\cdot)$ are set to zero on the right of t_0^k.

8.5   Design of the desired contact force during constraint phases

For the sake of simplicity we consider the case of the constraint phase Ω_k^J, $J \neq \emptyset$ with $J = \{1, \ldots, p\}$. Obviously a sufficiently large desired contact force P_d assures a constrained movement on Ω_k^J. Nevertheless at the end of the Ω_k^J phases a detachment from some surfaces Σ_i has to take place. It is clear that a take-off implies not only a well-defined desired trajectory but also some small values of the corresponding contact force components. On the other hand, if the components of the desired contact force decrease too much a detachment can take place before the end of the Ω_k^J phases which can generate other impacts. Therefore we need a lower bound of the desired force which assures the contact during the Ω_k^J phases. Dropping the time argument, the dynamics of the system on Ω_k^J can be written as

$$\begin{cases} M(q)\ddot{q} + F(q,\dot{q}) = U_c + D_p^T \lambda_p \\ \quad\quad 0 \leq q_p \perp \lambda_p \geq 0 \end{cases} \tag{8.15}$$

where $F(q,\dot{q}) = C(q,\dot{q})\dot{q} + G(q)$ and $D_p = [I_p \vdots O] \in \mathbb{R}^{p \times n}$. On Ω_k^J the system is permanently constrained which implies $q_p(\cdot) = 0$ and $\dot{q}_p(\cdot) = 0$. In order to assure these conditions it is sufficient to have $\lambda_p > 0$.

In the following let us denote $C(q,\dot{q}) = \begin{pmatrix} C(q,\dot{q})_{p,p} & C(q,\dot{q})_{p,n-p} \\ C(q,\dot{q})_{n-p,p} & C(q,\dot{q})_{n-p,n-p} \end{pmatrix}$ and

$$M^{-1}(q) = \begin{pmatrix} [M^{-1}(q)]_{p,p} & [M^{-1}(q)]_{p,n-p} \\ [M^{-1}(q)]_{n-p,p} & [M^{-1}(q)]_{n-p,n-p} \end{pmatrix} \text{ where the meaning of each}$$

component is obvious.

Proposition 8.2. *On* Ω_k^J *the constraint motion of the closed-loop system* *(8.15),(8.7),(8.9) is assured if the desired contact force is defined by*

$$(\lambda_d)_p \triangleq \beta - \frac{\bar{M}_{p,p}(q)}{1+K_f} \left([M^{-1}(q)]_{p,p} C_{p,n-p}(q,\dot{q}) + \right. \tag{8.16}$$
$$\left. + [M^{-1}(q)]_{p,n-p} C_{n-p,n-p}(q,\dot{q}) + \gamma_1 [M^{-1}(q)]_{p,n-p} \right) s_{n-p}$$

where $\bar{M}_{p,p}(q) = \left([M^{-1}(q)]_{p,p} \right)^{-1} = \left(D_p M^{-1}(q) D_p^T \right)^{-1}$ *is the inverse of* *the Delassus' matrix (see [Acary and Brogliato, 2008; Brogliato, 1999] for* *the definition) and* $\beta \in \mathbb{R}^p$, $\beta > 0$.

Remark 8.5. The control law used in this chapter with the design of λ_d described above leads to the following closed-loop dynamics on Ω_k^J.

$$\begin{cases} M_{p,n-p}(q)\dot{s}_{n-p} + C_{p,n-p}(q,\dot{q})s_{n-p} = (1+K_f)(\lambda - \lambda_d)_p \\ M_{n-p,n-p}(q)\dot{s}_{n-p} + C_{n-p,n-p}(q,\dot{q})s_{n-p} + \gamma_1 s_{n-p} = 0 \\ q_p = 0, \quad \lambda_p = \beta. \end{cases}$$

It is noteworthy that the closed-loop dynamics is nonlinear and therefore, we do not use the feedback stabilization proposed in [McClamroch and Wang, 1988].

8.6   Strategy for take-off at the end of constraint phases Ω_k^J

We have discussed in the previous sections the necessity of a trajectory with impacts in order to assure the robust stabilization on $\partial\Phi$ in finite time and, the design of the desired trajectory to stabilize the system on $\partial\Phi$. Now, we are interested in finding the conditions on the control signal U_c^J that assure the take-off at the end of the constrained phases Ω_k^J. We consider the phase Ω_k^J expressed as the time interval $[t_f^k, t_d^k)$. The dynamics on $[t_f^k, t_d^k)$ is given by (8.15) and the system is permanently constrained, which implies $q_p(\cdot) = 0$ and $\dot{q}_p(\cdot) = 0$. Let us also consider that the first r constraints $(r < p)$ have to be deactivated. Thus, the detachment takes place at t_d^k if $\ddot{q}_r(t_d^{k+}) > 0$ which requires $\lambda_r(t_d^{k-}) = 0$. The last $p - r$ constraints remain active which means $\lambda_{p-r}(t_d^{k-}) > 0$.

To simplify the notation we drop the arguments t and q in many equations of this section. We decompose the LCP matrix as:

$$(1 + K_f)D_p M^{-1}(q)D_p^\top = \begin{pmatrix} A_1(q) & A_2(q) \\ A_2(q)^T & A_3(q) \end{pmatrix}$$

with $A_1 \in \mathbb{R}^{r \times r}, A_2 \in \mathbb{R}^{r \times (p-r)}$ and $A_3 \in \mathbb{R}^{(p-r) \times (p-r)}$.

Proposition 8.3. *For the closed-loop system (8.15), (8.7), (8.9) the decrease of the active constraints number from p to $p - r$ (with $r < p$), is possible if*

$$\begin{pmatrix} (\lambda_d)_r \left(t_d^k\right) \\ (\lambda_d)_{p-r} \left(t_d^k\right) \end{pmatrix} = \begin{pmatrix} \left(A_1 - A_2 A_3^{-1} A_2^T\right)^{-1} \left(b_r - A_2 A_3^{-1} b_{p-r}\right) - C_1 \\ C_2 + A_3^{-1} \left(b_{p-r} - A_2^T (\lambda_d)_r\right) \end{pmatrix} \tag{8.17}$$

where

$$b_p \triangleq b(q, \dot{q}, U_{nc}) \triangleq D_p M^{-1}(q)[U_{nc} - F(q, \dot{q})] \geq 0$$

and $C_1 \in \mathbb{R}^r$, $C_2 \in \mathbb{R}^{p-r}$ such that $C_1 \geq 0$, $C_2 > 0$.

8.7   Closed-loop stability analysis

In the case $\Phi = \mathbb{R}^n$, the function $V(t, s, \tilde{q})$ in (8.10) can be used in order to prove the closed-loop stability of the system (8.8), (8.9) (see for instance [Brogliato *et al.*, 2007]. In the case studied here ($\Phi \subset \mathbb{R}^n$) the analysis becomes more complex.

To simplify the notation $V(t, s(t), \tilde{q}(t))$ is denoted as $V(t)$. In order to introduce the main result of this chapter we make the next assumption, which is verified in practice for dissipative systems.

Assumption 8.1. The controller U_t in (8.9) assures that all the transition phases are finite (see Definition 8.6) and the accumulation point t_∞^k is smaller than $t_d^{k,1}$ for all $k \in \mathbb{N}$.

Since outside $[t_0^k, t_f^k]$ we will show that the Lyapunov function exponentially decreases, we may presume that all the impacts take place during I_k.

Lemma 8.1. *Consider the closed-loop system (8.7)-(8.9) with $(q_d^*)_p(\cdot)$ defined on the interval $[\tau_0^k, t_0^k]$ as in (8.12)-(8.11). Let us also suppose that* *condition* **b)** *of Proposition 8.1 is satisfied. The following inequalities hold:*

$$\|\tilde{q}(t_0^{k-})\| \leq \sqrt{\frac{V(\tau_0^k)}{\gamma_1\gamma_2}}, \ \|s(t_0^{k-})\| \leq \sqrt{\frac{2V(\tau_0^k)}{\lambda_{min}(M(q))}}$$

$$\|\dot{\tilde{q}}(t_0^{k-})\| \leq \left(\sqrt{\frac{2}{\lambda_{min}(M(q))}} + \sqrt{\frac{\gamma_2}{\gamma_1}}\right) V^{1/2}(\tau_0^k)$$

(8.18)

Furthermore, if $t_0^k \leq \tau_1^k$ *and* t_0^k *is a* p_{ϵ_k}*-impact one has*

$$\|(q_d)_p(t_0^{k-})\| \leq \epsilon_k + \sqrt{\frac{V(\tau_0^k)}{\gamma_1\gamma_2}}$$

$$\|(\dot{q}_d)_p(t_0^{k-})\| \leq K + K'V^{1/3}(\tau_0^k)$$

(8.19)

where $\epsilon_k \leq \max\{\psi, \sqrt{p}\varphi V^{1/3}(\tau_0^k) + \sqrt{\frac{V(\tau_0^k)}{\gamma_1\gamma_2}}\}$, $K = \frac{6p\psi}{\tau_1^k - \tau_0^k - \delta}$ *and*

$$K' = \frac{6\sqrt{p}}{\tau_1^k - \tau_0^k - \delta}\sqrt{\left(\frac{2}{\sqrt{\gamma_1\gamma_2}} + (1 + \sqrt{p})\varphi\right)(\varphi + \psi) + \psi\varphi}$$

The main result of this chapter can be stated as follows.

Theorem 8.1. *Let Assumption 1 hold,* $e \in [0, 1)$ *and* $(q_d^*)_p(\cdot)$ *defined as in (8.12)-(8.11). The closed-loop system (8.7)-(8.9) initialized on* Ω_0 *such that* $V(\tau_0^0) \leq 1$, *satisfies the requirements of Proposition 8.1 and is therefore practically weakly stable with the closed-loop state* $x(\cdot) = [s(\cdot), \tilde{q}(\cdot)]$ *and* $R = \sqrt{e^{-\gamma(t_f^k - t_\infty^k)}(1 + K_1 + K_2 + \xi)/\rho}$ *where* $\rho = \min\{\lambda_{min}(M(q))/2; \gamma_1\gamma_2\}$

$$K_1 = \sqrt{3}\gamma_1\gamma_2\varphi(\psi + \sqrt{p}\varphi + \frac{1}{\sqrt{\gamma_1\gamma2}})$$

$$K_2 = \lambda_{max}(M(q))\left[3KK' + \frac{3}{2}(K')^2 + \gamma_2\psi K + \sqrt{\frac{2\gamma_2}{\lambda_{min}(M(q))\gamma_1}} + \frac{4\gamma_2}{\gamma_1}\right.$$

$$+ (K' + K)\left(\gamma_2\sqrt{p}\varphi + 3\sqrt{\frac{\gamma_2}{\gamma_1}} + \sqrt{\frac{2}{\lambda_{min}(M(q))}}\right) + \gamma_2^2\psi\varphi\sqrt{p}$$

$$+ \gamma_2\left(\sqrt{\frac{2}{\lambda_{min}(M(q))}} + 3\sqrt{\frac{\gamma_2}{\gamma_1}}\right)(\psi + \sqrt{p}\varphi) + \frac{\gamma_2^2\varphi^2 p}{2}$$

$$+ \psi\gamma_2\left(2\sqrt{\frac{\gamma_2}{\gamma_1}} + \sqrt{\frac{2}{\lambda_{min}(M(q))}} + K'\right)\right]$$

Remark 8.6. Since the closed-loop system (8.7)-(8.9) satisfies the requirements of Proposition 8.1 one also deduces $V(\tau_0^k) \leq \delta(\gamma, \xi)$, so $\epsilon_k \leq \max\{\psi, \sqrt{\bar{p}}\varphi\delta(\gamma, \xi)^{1/3} + \sqrt{\frac{\delta(\gamma, \xi)}{\gamma_1\gamma_2}}\}$, $\forall k \geq 1$. In other words the sequence $\{\epsilon_k\}_k$ is uniformly upperbounded and the upperbound can be decreased by adjusting the parameters ψ and γ.

8.8   Illustrative example

Let us consider a two-link planar manipulator with admissible domain $\Phi = \{(x, y) \mid y \geq 0, 0.7 - x \geq 0\}$. Let us also consider an unconstrained desired trajectory given by the circle $\{(x, y) \mid (x - 0.7)^2 + y^2 = 0.5\}$ that violates both constraints. In other words, the two-link planar manipulator must track a quarter-circle; stabilize on and then follow the line $\Sigma_1 = \{(x, y) \mid y = 0\}$; stabilize on the intersection of Σ_1 and $\Sigma_2 = \{(x, y) \mid x = 0.7\}$; detach from Σ_1 and follow Σ_2 until the unconstrained circle re-enters Φ and finally take-off from Σ_2 in order to repeat the previous steps. Therefore, we have: $\mathbb{R}^+ = \Omega_0^{\emptyset} \cup I_1 \cup \Omega_1^{J_{1,1}} \cup I_2 \cup \Omega_2^{J_{2,1}} \cup \Omega_2^{J_{2,2}} \cup \Omega_2^{J_{2,3}} \cup I_3 \cup \Omega_3^{J_{3,1}} \cup I_4 \cup ...$ with $J_{1,1} = \{1\}$, $J_{2,1} = \{1, 2\}$, $J_{2,2} = \{2\}$, $J_{2,3} = \emptyset$, etc. We note that during I_{2k+1} the system is stabilized on Σ_1 (1-impacts) while during I_{2k} the system is stabilized on $\Sigma_1 \cap \Sigma_2$ (2_{ϵ_k} −impacts). We define task1= $\Omega_0^{\emptyset} \cup I_1 \cup \Omega_1^{J_{1,1}} \cup I_2 \cup \Omega_2^{J_{2,1}} \cup \Omega_2^{J_{2,2}}$ and task2= $\Omega_2^{J_{2,3}} \cup I_3 \cup \Omega_3^{J_{3,1}} \cup I_4 \cup \Omega_4^{J_{4,1}} \cup \Omega_4^{J_{4,2}}$. Each task actually corresponds to a complete tracking of the quarter-circle.

The numerical values used for the dynamical model are again $l_1 = l_2 = 0.5m$, $I_1 = I_2 = 1kg.m^2$, $m_1 = m_2 = 1kg$ and the restitution coefficient $e = 0.7$. The impacts are imposed by $\varphi = 100$ in (8.12) (8.13) and the beginning of transition phases are defined using $\psi = 0.05$ in (8.14). We impose a period of 10 seconds for two consecutive cycles and we simulate the dynamics during 60 seconds. The simulations have been performed with the SICONOS platform available at: http://siconos.gforge.inria.fr/.

To evaluate the performances of the controller, two criteria will be used: $C_1 = \max_{t\in\Delta} |T_2(q(t))U(t)|$ and $C_2 = \int_\Delta ||\tilde{q}(t)||^2 + ||\dot{\tilde{q}}(t)||^2 dt$ where $\Delta = [0, t_d^{2,2}]$ for task1 and $\Delta = [t_d^{2,2}, t_d^{4,2}]$ for task2.

From Figures 8.5-8.8 one deduces that the best performances in terms of both criteria are obtained when γ_2 is small (around 10) and $\gamma_1\gamma_2$ is around 80. However, if γ_2 is too small then Assumption 8.1 is violated. It is noteworthy that the both criteria significantly decrease from task1 to task2.

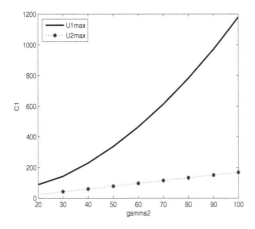

Fig. 8.5 $C_1(\gamma_2)$ on task1 (solid) and task2 (dashed) with $\gamma_1 = 25$.

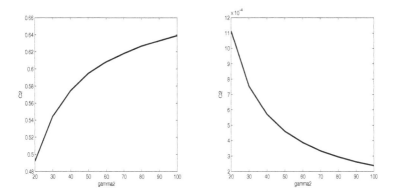

Fig. 8.6   Left: $C_2(\gamma_2)$ on task1; Right: $C_2(\gamma_2)$ on task2 with $\gamma_1 = 25$.

8.9   Conclusions

In this chapter we have proposed a methodology to study the tracking control of fully actuated Lagrangian systems subject to multiple frictionless unilateral constraints and multiple impacts. The main contribution of the work is twofold: first, it formulates a general control framework and second, it provides a stability analysis for the class of systems under consideration.

C. Morarescu, B. Brogliato, T. Nguyen

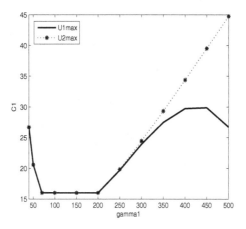

Fig. 8.7 $C_1(\gamma_1)$ on task1 (solid) and task2 (dashed) with $\gamma_1\gamma_2 = 800$.

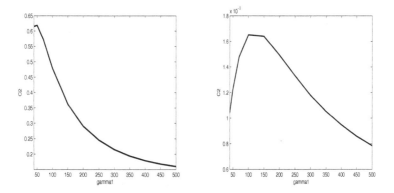

Fig. 8.8   Left: $C_2(\gamma_1)$ on task1; Right: $C_2(\gamma_1)$ on task2 with $\gamma_1\gamma_2 = 800$.

It is noteworthy that even in the simplest case of only one frictionless unilateral constraint the chapter already presents some notable improvements
with respect to the existing works. Precisely, the stability analysis result is
significantly more general than those presented in [Bourgeot and Brogliato,
2005] and [Brogliato, 1999] and, each element entering the dynamics (desired trajectory, contact force) is explicitly defined. Numerical simulations
are done with the SICONOS software platform [Acary and Brogliato, 2008]

in order to illustrate the results.

References

Acary, V. and Brogliato, B. (2008), *Numerical Methods for Nonsmooth Dynamical Systems*, Lecture Notes in Applied and Computational Mechanics, vol **35**, Springer-Verlag, Berlin Heidelberg.

Acary, V. and Pérignon, F. (2007), *An introduction to* SICONOS, INRIA Technical report RT 340, SICONOS open-source software available at: http://siconos.gforge.inria.fr/.

Ballard, P. (2001), *Formulation and well-posedness of the dynamics of rigid body systems with perfect unilateral constraints*, Phil. Trans. R. Soc. Lond. **A359**, 2327-2346.

Bentsman, J. and Miller, B. M. (2007), *Dynamical systems with active singularities of elastic type: A modeling and controller synthesis framework*, IEEE Transactions on Automatic Control, vol.**52**, No.1, 39-55.

Bourgeot, J.-M. and Brogliato, B. (2005),*Tracking control of complementarity lagrangian systems*, International Journal of Bifurcation and Chaos, **15**(6), 1839-1866.

Brogliato, B. (1999), *Nonsmooth Mechanics*, Springer CCES, London, 2nd Ed.

Brogliato, B. (2004), *Absolute stability and the Lagrange-Dirichlet theorem with monotone multivalued mappings*, Systems and Control Letters, vol.**51**, 343-353.

Brogliato, B., Lozano, R., Maschke, B. and Egeland, O. (2007), *Dissipative Systems Analysis and Control. Theory and Applications*, Springer CCES, London, 2nd Ed.

Brogliato, B., Niculescu, S.-I., Orhant, P. (1997), *On the control of finite-dimensional mechanical systems with unilateral constraints*, IEEE Trans. Autom. Contr. **42**(2), 200-215.

Facchinei, F. and Pang, J.S. (2003), *Finite-dimensional Variational Inequalities and Complementarity Problems*, Springer Series in Operations Research, New-York.

Faulring, E.L. and Lynch, K.M. and Colgate, J.E. and Peshkin, M.A (2007), *Haptic display of constrained dynamic systems via admittance displays*, IEEE Transactions on Robotics, vol.**23**, No.1, 101-111.

Galeani, S., Menini, L., Potini, A. and Tornambè, A. (2008), *Trajectory tracking for a particle in elliptical billiards*, International Journal of Control, vol.**81**(2), 189-213.

Glocker, C. (2001), *Set Value Force Laws: Dynamics of Non-Smooth Systems* Lecture Notes in Applied Mechanics, Vol.1, Springer.

Glocker, C. (2004), *Concepts for Modeling Impacts without Friction*, Acta Mechanica, Vol. **168**, 1-19.

Huang, H. P. and McClamroch, N. H. (1988), *Time optimal control for a robotic contour following problem*, IEEE J. Robotics and Automation, Vol. **4**, 140-

149.

Lee, E. and Park, J. and Loparo, K.A. and Schrader, C.B. and Chang, P.H. (2003), *Bang-bang impact control using hybrid impedance/time-delay control*, IEEE/ASME Transactions on Mechatronics, vol.**8**, No.2, 272-277.

Leine, R.I. and van de Wouw, N. (2008), *Stability properties of equilibrium sets of nonlinear mechanical systems with dry friction and impact*, Nonlinear Dynamics, vol.**51**(4), 551-583.

Leine, R.I. and van de Wouw, N. (2008), *Uniform convergence of monotone measure differential inclusions: with application to the control of mechanical systems with unilateral constraints*, Int. Journal of Bifurcation and Chaos, vol.**18** (5), 1435-1457.

Leine, R.I. and van de Wouw, N. (2008), *Stability and Convergence of Mechanical Systems with Unilateral Constraints*, Lecture Notes in Applied and Computational Mechanics, vol.36, Springer Verlag Berlin Heidelberg.

Hiriart-Urruty J.B. and Lemaréchal, C. (2001), *Fundamentals of Convex Analysis*, Springer-Verlag, Grundlehren Texts in Mathematics.

Mabrouk, M. (1998), *A unified variational model for the dynamics of perfect unilateral constraints*, European Journal of Mechanics A/Solids, **17**(5), 819-842.

McClamroch, N.H. and Wang, D. (1988), *Feedback stabilization and tracking of constrained robots*, IEEE Trans. Autom. Contr. **33**, 419-426.

Menini, L. and Tornambè, A. (2001), *Asymptotic tracking of periodic trajectories for a simple mechanical system subject to nonsmooth impacts*, IEEE Trans. Autom. Contr. **46**, 1122-1126.

Menini, L. and Tornambè, A: (2001), *Exponential and BIBS stabilisation of one degree of freedom mechanical system subject to single non-smooth impacts*, IEE Proc.- Control Theory Appl. Vol.**148**, No. 2, 147-155.

Menini, L. and Tornambè, A: (2001), *Dynamic position feedback stabilisation of multidegrees-of-freedom linear mechanical system subject to nonsmooth impacts*, IEE Proc.- Control Theory Appl. Vol.**148**, No. 6, 488-496.

Moreau, J. J (1988), *Unilateral contact and dry friction in finite freedom dynamics*, Nonsmooth Mechanics and Applications, CISM Courses and Lectures, Vol. **302** (Springer-Verlag).

Pagilla, P.R. (2001), *Control of contact problem in constrained Euler-Lagrange cystems*, IEEE Transactions on Automatic Control, vol.**46**, No. 10, 1595-1599.

Pagilla, P.R. and Yu, B. (2004), *An Experimental Study of Planar Impact of a Robot Manipulator*, IEEE/ASME Transactions on Mechatronics, vol.**9**, No. 1, 123-128.

Sekhavat, P. and Sepehri, N. and Wu, Q. (2006), *Impact stabilizing controller for hydraulic actuators with friction: Theory and experiments*, Control Engineering Practice, vol.**14**, 1423-1433.

Slotine, J. J. and Li, W. (1988), *Adaptive manipulator control: A case study*, IEEE Trans. Automatic Control, vol.**33**, 995-1003.

Tornambé, A. (1999), *Modeling and control of impact in mechanical systems: Theory and experimental results*, IEEE Transactions on Automatic Control, vol.**44**, No. 2, 294-309.

van Vliet, J. and Sharf, I. and Ma, O. (2000), *Experimental validation of contact dynamics simulation of constrained robotic tasks*, International Journal of Robotics Research, vol.**19**, No.12, 1203-1217.

Volpe, R. and and Khosla, P. (1993), *A theoretical and experimental investigation of impact control for manipulators*, International Journal of Robotics Research, vol.**12**, No.4, 351–365.

Xu, W.L.and Han, J.D. and Tso, S.K. (2000), *Experimental study of contact transition control incorporating joint acceleration feedback*, IEEE/ASME Transactions on Mechatronics, vol.**5**, No.3, 292-301.

Stability and Control of Lur'e-type Measure Differential Inclusions

Nathan van de Wouw*, Remco I. Leine†

*Department of Mechanical Engineering,
Eindhoven University of Technology,
5600 MB, Eindhoven, The Netherlands

†Institute of Mechanical Systems,
Department of Mechanical and Process Engineering,
ETH Zurich, CH-8092 Zurich, Switzerland

Abstract

In this paper we present results on, firstly, the stability analysis for perturbed Lur'e-type measure differential inclusions and, secondly, the tracking control problem for this class of systems. The framework of measure differential inclusions allows us to describe systems with discontinuities in the state evolution, such as for example mechanical systems with unilateral constraints. As a stepping stone, we present results on the stability of time-varying solutions of such systems in the scope of the convergence property. Next, this property is exploited to provide a solution to the tracking problem. The results are illustrated by application to a mechanical motion system with a unilateral velocity constraint.

9.1 Introduction

In this paper, we, firstly, study the stability of time-varying solutions of Lur'e-type measure differential inclusions and, secondly, provide a solution to the tracking problem for Lur'e-type measure differential inclusions.

The mathematical formalism of measure differential inclusions (MDIs) can be used to describe systems with non-smooth and even impulsive dynamics. In Lur'e-type MDIs, the system can be represented as a feedback interconnection of a smooth linear dynamic part and a static discontinuous (and possibly impulsive) nonlinearity. This model class can be used to represent a wide range of engineering systems, such as e.g. mechanical (motion) systems with friction, unilateral contact and impact or electronic systems with switches and diodes.

Non-smooth dynamical systems, with or without impulsive dynamics, are studied by various scientific communities using different mathematical frameworks [Leine *et. al.*, 2004] : singular perturbations, switched or hybrid systems, complementarity systems, (measure) differential inclusions. The singular perturbation approach replaces the non-smooth system by a singularly perturbed smooth system. The resulting ordinary differential equation is extremely stiff and hardly suited for numerical integration. In the field of systems and control theory, the term hybrid system is frequently used for systems composed of continuous differential equations and discrete-event parts [Brogliato, 1999; van der Schaft and Schumacher, 2000; Goebel *et. al.*, 2009]. The switched or hybrid system concept switches between differential equations with possible state re-initialisations and is not able to describe accumulation points, e.g. infinitely many switching events which occur in a finite time such as a bouncing ball coming to rest on a table, in the sense that solutions can not proceed over the accumulation point. Systems described by differential equations with a discontinuous right-hand side, but with a time-continuous state, can be extended to differential inclusions with a set-valued right-hand side [Filippov, 1988]. The differential inclusion concept gives a simultaneous description of the dynamics in terms of a single inclusion, which avoids the need to switch between different differential equations. Moreover, this framework is able to describe accumulation points of switching events.

Measure differential inclusions can be used to describe systems which expose discontinuities in the state and/or vector field [Monteiro Marques, 1993; Moreau, 1988a; Brogliato, 1999; Acary *et. al.*, 2008]. The differential measure of the state vector does not only consist of a part with a density with respect to the Lebesgue measure (i.e. the time-derivative of the state vector), but is also allowed to contain an atomic part. The dynamics of the system is described by an inclusion of the differential measure of the state to a state-dependent set (similar to the concept of differential inclusions). Consequently, the measure differential inclusion concept describes the continuous dynamics as well as the impulsive dynamics with a single statement in terms of an inclusion and is able to describe accumulation phenomena. An advantage of this framework over other frameworks, such as the hybrid systems formalism [van der Schaft and Schumacher 2000; Goebel *et. al.*, 2009], is the fact that physical interaction laws, such as friction and impact in mechanics or diode characteristics in electronics, can be formulated as set-valued force laws and be seamlessly incorporated in the formulation, see e.g. [Glocker, 2001; Leine and van de Wouw, 2008].

Stability properties of measure differential inclusions are essential in both bifurcation analysis and the control of such systems. In [Leine and van de Wouw, 2008], results on the stability of stationary sets of measure differential inclusions (with a special focus on mechanical systems with unilateral constraints) are presented. In [Brogliato, 2004], stability properties of an equilibrium of measure differential inclusions of Lur'e-type are studied. The nonlinearities in the feedback loop are required to exhibit monotonicity properties and, if additionally passivity conditions on the linear part of the system are assured, then stability of the equilibrium can be guaranteed. Note that this work studies the stability of stationary solutions. However, many control problems, such as tracking control, output regulation, synchronisation and observer design require the stability analysis of time-varying solutions. The research on the stability and stabilisation of time-varying solutions of non-smooth systems (especially with state jumps) is still in its infancy and the current paper should be placed in this context. It should be noted that the tracking control of measure differential inclusions has received very little attention in literature, see [Bourgeot and Brogliato, 2005; Brogliato *et. al.*, 1997; Menini and Tornambè, 2001] for works focusing on mechanical systems with unilateral constraints. Recent work on the observer design for Lur'e-type systems with set-valued nonlinearities in the feedback loop can be found in [Brogliato and Heemels, 2009].

In order to study the stability of certain time-varying solutions we consider the framework of convergence. A system, which is excited by an input, is called convergent if it has a unique solution that is bounded on the whole time axis and this solution is globally asymptotically stable. Obviously, if such a solution does exist, then all other solutions converge to this solution, regardless of their initial conditions, and can be considered as a steady-state solution [Demidovich, 1967; Pavlov et. al., 2005; Pavlov et. al., 2004]. Similar notions describing the property of solutions converging to each other are studied in literature. The notion of contraction has been introduced in [Lohmiller and Slotine, 1998] (see also references therein). An operator-based approach towards studying the property that all solutions of a system converge to each other is pursued in [Fromion et al., 1996]. In [Angeli, 2002], a Lyapunov approach has been developed to study the global uniform asymptotic stability of all solutions of a system (in [Angeli, 2002], this property is called incremental stability).

The convergence property of a system plays an important role in many nonlinear control problems including tracking, synchronization, observer design, and the output regulation problem, see e.g. [Pavlov et. al., 2005; Pogromsky, 1998, van de Wouw and Pavlov, 2008] and references therein. Namely, in many control problems, such as the tracking problem, it is required that controllers are designed in such a way that all solutions of the corresponding closed-loop system "forget" their initial conditions. Actually, one of the main tasks of feedback is to eliminate the dependency of solutions on initial conditions. In this case, all solutions converge to some steady-state solution that is determined only by the input of the closed-loop system. This input can be, for example, a command signal or a signal generated by a feedforward part of the controller or, as in the observer design problem, it can be the measured signal from the observed system. Secondly, from a dynamics point of view, convergence is an interesting property because it excludes the possibility of different coexisting steady-state solutions: namely, a convergent system excited by a periodic input has a *unique* globally asymptotically stable periodic solution. Moreover, the notion of convergence is a powerful tool for the analysis of time-varying systems. This tool can be used, for example, for performance analysis of nonlinear control systems, see e.g. [van de Wouw et. al., 2008].

In [Demidovich, 1967], conditions for the convergence property were formulated for smooth nonlinear systems. In [Yakubovich, 1964], Lur'e-type systems, possibly with discontinuities, were considered and convergence conditions proposed. We note that in the work of [Yakubovich, 1964] no impulsive right-hand sides were considered. Sufficient conditions for both continuous and discontinuous piece-wise affine (PWA) systems have been proposed in [Pavlov et. al., 2007]. Recently, also convergence conditions for complementarity systems [Camlibel and van de Wouw, 2007] and nonlinear discrete-time systems [Pavlov and van de Wouw, 2008] have been proposed.

In this paper, we will provide sufficient conditions for the convergence property of Lur'e-type measure differential inclusions, i.e. measure differential inclusions consisting of a linear plant and output-dependent set-valued nonlinearities in the feedback loop, and will exploit this property to tackle the tracking problem for such systems.

The outline of the paper is as follows. Section 9.2 treats some mathematical preliminaries regarding properties of set-valued functions and passive systems, which are needed in the remainder of this paper. Section 9.3 provides a brief introduction to measure differential inclusions. Subsequently, we define the convergence property of dynamical systems in Section 9.4. In Section 9.5, we present sufficient conditions for the convergence of Lur'e-type measure differential inclusions. These results on convergence are exploited in Section 9.6 to provide a solution to the tracking control problem. An illustrative example of a mechanical system with a unilateral constraint is discussed in detail in Section 9.7. Finally, Section 9.8 presents concluding remarks.

9.2   Preliminaries

We first define what we mean by a set-valued function.

Definition 9.1 (Set-valued Function). *A set-valued function* $\mathcal{F}$: $\mathbb{R}^n \to \mathbb{R}^n$ *is a map that associates with any* $\boldsymbol{x} \in \mathbb{R}^n$ *a set* $\mathcal{F}(\boldsymbol{x}) \subset \mathbb{R}^n$.

A set-valued function can therefore contain vertical segments on its graph denoted by Graph($\mathcal{F}$). Here, we define the graph of a (set-valued) function $\mathcal{F}(\cdot)$ as Graph($\mathcal{F}$) = $\{(\boldsymbol{x}, \boldsymbol{y}) | \boldsymbol{y} \in \mathcal{F}(\boldsymbol{x})\}$. We use the graph to define monotonicity of a set-valued function [Aubin and Frankowska, 1990]. Next let us define the concept of a *maximal monotone* set-valued function.

Definition 9.2 (Maximal Monotone Set-valued Function).
A set-valued function $\mathcal{F}(\boldsymbol{x})$: $\mathbb{R}^n \to \mathbb{R}^n$ *is called monotone if its graph is monotone in the sense that for all* $(\boldsymbol{x}, \boldsymbol{y}) \in Graph(\mathcal{F})$ *and for all* $(\boldsymbol{x}^*, \boldsymbol{y}^*) \in Graph(\mathcal{F})$ *it holds that* $(\boldsymbol{y} - \boldsymbol{y}^*)^{\mathrm{T}}(\boldsymbol{x} - \boldsymbol{x}^*) \geq 0$. *In addition, if* $(\boldsymbol{y} - \boldsymbol{y}^*)^{\mathrm{T}}(\boldsymbol{x} - \boldsymbol{x}^*) \geq \alpha \|\boldsymbol{x} - \boldsymbol{x}^*\|^2$ *for some* $\alpha > 0$, *then the set-valued map is strictly monotone. A monotone set-valued function* $\mathcal{F}(\boldsymbol{x})$ *is called maximal monotone if there exists no other monotone set-valued function whose graph strictly contains the graph of* $\mathcal{F}$. *If* $\mathcal{F}$ *is strictly monotone*

and maximal, then it is called strictly maximal monotone.

In Section 9.4 we will exploit the concept of passivity of a linear time-invariant system, which we define below.

Definition 9.3. The system $\dot{x} = Ax + Bu$, $y = Cx$ or the triple (A, B, C) is said to be strictly passive if there exist an $\varepsilon > 0$ and a matrix $P = P^{\mathrm{T}} > 0$ such that

$$A^{\mathrm{T}}P + PA \leq -\varepsilon I, \quad B^{\mathrm{T}}P = C. \tag{9.1}$$

9.3 Measure differential inclusions

In this section, we introduce the measure differential inclusion

$$\mathrm{d}x \in \mathrm{d}\boldsymbol{\Gamma}(t, x(t)) \tag{9.2}$$

as has been proposed by Moreau (1988). The concept of differential inclusions has been extended to measure differential inclusions in order to allow for discontinuities in the evolution of the state $x(t)$, see e.g. [Monteiro Marques, 1993; Moreau, 1988; Brogliato, 1999]. With the differential inclusion $\dot{x}(t) \in \mathcal{F}(t, x(t))$, in which $\mathcal{F}(t, x(t))$ is a set-valued mapping, we are able to describe a non-smooth absolutely continuous time-evolution $x(t)$. The solution $x(t) : \mathcal{I} \to \mathbb{R}^n$ fulfills the differential inclusion almost everywhere, because $\dot{x}(t)$ does not exist on a Lebesgue negligible set $\mathcal{D} \subset \mathcal{I}$ of time-instances $t_i \in \mathcal{D}$ related to non-smooth state evolution. Instead of using the density $\dot{x}(t)$, we can also write the differential inclusion using the differential measure:

$$\mathrm{d}x \in \mathcal{F}(t, x(t))\,\mathrm{d}t, \tag{9.3}$$

which yields a measure differential inclusion (with $\mathrm{d}t$ the Lebesgue measure). The solution $x(t)$ fulfills the measure differential inclusion (9.3) *for all* $t \in I$ because of the underlying integration process being associated with measures. Moreover, writing the dynamics in terms of a measure differential inclusion allows us to study a larger class of functions $x(t)$, as we can let the differential measure of the state $\mathrm{d}x$ contain parts other than the Lebesgue integrable part. In order to describe a time-evolution of bounded variation which is discontinuous at isolated time-instances, we let the differential measure $\mathrm{d}x$ also have an atomic part:

$$\mathrm{d}x = \dot{x}(t)\,\mathrm{d}t + (x^+(t) - x^-(t))\,\mathrm{d}\eta, \tag{9.4}$$

where $\mathrm{d}\eta$ is the atomic differential measure, being the sum of Dirac point measures, as defined in [Glocker, 2001; Leine and van de Wouw, 2008], and $\boldsymbol{x}^+(t) = \lim_{\tau\downarrow 0}\boldsymbol{x}(t+\tau)$, $\boldsymbol{x}^-(t) = \lim_{\tau\uparrow 0}\boldsymbol{x}(t+\tau)$. Therefore, we extend the measure differential inclusion (9.3) with an atomic part as well:

$$\mathrm{d}\boldsymbol{x} \in \mathcal{F}(t, \boldsymbol{x}(t))\,\mathrm{d}t + \mathcal{G}(t, \boldsymbol{x}^-(t), \boldsymbol{x}^+(t))\,\mathrm{d}\eta.$$

Here, $\mathcal{G}(t, \boldsymbol{x}^-(t), \boldsymbol{x}^+(t))$ is a set-valued mapping, which in general depends on t, $\boldsymbol{x}^-(t)$ and $\boldsymbol{x}^+(t)$. More conveniently, and with some abuse of notation, we write the measure differential inclusion as in (9.2), where $\mathrm{d}\boldsymbol{\Gamma}(t, \boldsymbol{x}(t))$ is a set-valued measure function defined as

$$\mathrm{d}\boldsymbol{\Gamma}(t, \boldsymbol{x}(t)) = \mathcal{F}(t, \boldsymbol{x}(t))\,\mathrm{d}t + \mathcal{G}(t, \boldsymbol{x}^-(t), \boldsymbol{x}^+(t))\,\mathrm{d}\eta. \qquad (9.5)$$

The measure differential inclusion (9.2) has to be understood in the sense of integration and its solution $\boldsymbol{x}(t)$ is a function of locally bounded variation which fulfills $\boldsymbol{x}^+(t) = \boldsymbol{x}^-(t_0) + \int_I \boldsymbol{f}(t, \boldsymbol{x})\,\mathrm{d}t + \boldsymbol{g}(t, \boldsymbol{x}^-, \boldsymbol{x}^+)\,\mathrm{d}\eta$, for every compact interval $I = [t_0, t]$, where the functions $\boldsymbol{f}(t, \boldsymbol{x})$ and $\boldsymbol{g}(t, \boldsymbol{x}^-, \boldsymbol{x}^+)$ have to obey $\boldsymbol{f}(t, \boldsymbol{x}) \in \mathcal{F}(t, \boldsymbol{x})$, $\boldsymbol{g}(t, \boldsymbol{x}^-, \boldsymbol{x}^+) \in \mathcal{G}(t, \boldsymbol{x}^-(t), \boldsymbol{x}^+(t))$. Note that for functions of locally bounded variation, the limits defining $\boldsymbol{x}^+$ and $\boldsymbol{x}^-$ exist. Substitution of (9.4) in the measure differential inclusion (9.2), (9.5) gives

$$\dot{\boldsymbol{x}}(t)\,\mathrm{d}t + (\boldsymbol{x}^+(t) - \boldsymbol{x}^-(t))\,\mathrm{d}\eta \in \mathcal{F}(t, \boldsymbol{x}(t))\,\mathrm{d}t + \mathcal{G}(t, \boldsymbol{x}^-(t), \boldsymbol{x}^+(t))\,\mathrm{d}\eta,$$

which we can separate in the Lebesgue integrable part

$$\dot{\boldsymbol{x}}(t)\,\mathrm{d}t \in \mathcal{F}(t, \boldsymbol{x}(t))\,\mathrm{d}t,$$

and atomic part

$$(\boldsymbol{x}^+(t) - \boldsymbol{x}^-(t))\,\mathrm{d}\eta \in \mathcal{G}(t, \boldsymbol{x}^-(t), \boldsymbol{x}^+(t))\,\mathrm{d}\eta$$

from which we can retrieve

$$\dot{\boldsymbol{x}}(t) \in \mathcal{F}(t, \boldsymbol{x}(t))$$

and the jump condition

$$\boldsymbol{x}^+(t) - \boldsymbol{x}^-(t) \in \mathcal{G}(t, \boldsymbol{x}^-(t), \boldsymbol{x}^+(t)).$$

The latter formulation hints towards the relation with hybrid systems (or hybrid inclusions) as e.g. in [Goebel *et. al.*, 2009]. We note that here the above jump condition may generally be implicit in the sense that the map $\mathcal{G}(t, \boldsymbol{x}^-(t), \boldsymbol{x}^+(t))$ actually depends not only on t, $\boldsymbol{x}^-(t)$ but also on $\boldsymbol{x}^+(t)$. Such an implicit description of the post-jump state makes this formalism especially useful for the description of physical processes with set-valued reset laws, such as mechanical systems with unilateral constraints and electrical systems with set-valued elements (spark plugs, diodes and the like) [Glocker, 2005]. In mechanical systems with e.g. inelastic impacts the map $\mathcal{G}$ may only depend on t and $\boldsymbol{x}^+(t)$. The solution of the post-jump state constitutes a combinatorial problem which is inherent to the physical nature of unilateral constraints. The implicit description of the post-jump state is the key difference between the measure differential inclusion formalism and the hybrid system formalism, which pre-supposes an explicit jump map. Moreover, a description in terms of differential measures allows to describe accumulation points as an intrinsic part of the dynamics and also opens the way to the numerical treatment of systems with accumulation points.

It should be noted that the state $\boldsymbol{x}$ of (9.2) may be confined to a so-called admissible set, which we denote by $\mathcal{X}$. Here, we will assume that the measure differential inclusions under study exhibit the consistency property.

Definition 9.4 (Leine and van de Wouw, 2008). *The measure differential inclusion (9.2) is consistent if for any initial condition taken in its admissible set $\mathcal{X}$, i.e. $\boldsymbol{x}_0 = \boldsymbol{x}(t_0)$ is such that $\boldsymbol{x}_0 \in \mathcal{X}$, there exists a solution in forward time that resides in the admissible domain, i.e. $\boldsymbol{x}(t) \in \mathcal{X}$ for almost all $t \geq t_0$.*

9.4 Convergent systems

In this section, we briefly discuss the definition of convergence. Herein, the Lyapunov stability of solutions of (9.2) plays a central role. For the definition of stability of time-varying solutions we refer to [Willems, 1970; Pavlov et. al., 2005], or to [Leine and van de Wouw, 2008] for the specific case of measure differential inclusions. The definitions of convergence properties presented here extend the definition given in [Demidovich, 1967].

We consider systems of the form

$$d\boldsymbol{x} \in \mathcal{F}(\boldsymbol{x}(t), \boldsymbol{k}(t))\, dt + \mathcal{G}(\boldsymbol{x}^-(t), \boldsymbol{x}^+(t), \boldsymbol{K}(t))\, d\eta, \qquad (9.6)$$

with state $\boldsymbol{x} \in \mathbb{R}^n$ and where $\boldsymbol{k}(t)$, $\boldsymbol{K}(t) \in \mathbb{R}^d$ represent the non-impulsive and impulsive parts of the input, respectively. The function $\mathcal{G}(\boldsymbol{x}^-(t), \boldsymbol{x}^+(t), \boldsymbol{K}(t))$ is assumed to be affine in $\boldsymbol{K}(t)$. In the following, we will consider the inputs $\boldsymbol{k}(t) : \mathbb{R} \to \mathbb{R}^d$ to be in the class $\overline{\mathbb{PC}_d}$ of piecewise continuous inputs which are bounded on $\mathbb{R}$. Moreover, we will assume that $\boldsymbol{K}(t) : \mathbb{R} \to \mathbb{R}^d$ is zero almost everywhere, such that the impulsive inputs are separated in time, and bounded on $\mathbb{R}$; this class of functions will be denoted by $\boldsymbol{K}(t) \in \mathbb{M}_d$.

Let us formally define the property of convergence.

Definition 9.5. System (9.6) is said to be

- *exponentially convergent* if, for every input $\boldsymbol{k} \in \overline{\mathbb{PC}}_d$, $\boldsymbol{K} \in \mathbb{M}_d$, there exists a solution $\bar{\boldsymbol{x}}_k(t)$ satisfying the following conditions:

  (i) $\bar{\boldsymbol{x}}_k(t)$ is defined for almost all $t \in \mathbb{R}$
  (ii) $\bar{\boldsymbol{x}}_k(t)$ is bounded for all $t \in \mathbb{R}$ for which it is defined,
  (iii) $\bar{\boldsymbol{x}}_k(t)$ is globally exponentially stable.

The solution $\bar{\boldsymbol{x}}_k(t)$ is called a *steady-state solution* (where the subscript emphasizes the fact that the steady-state solution depends on the input, characterised by $\boldsymbol{k}(t)$ and $\boldsymbol{K}(t)$). As follows from the definition of convergence, any solution of a convergent system "forgets" its initial condition and converges to some steady-state solution. For *exponentially* convergent systems the steady-state solution is unique, as formulated below.

Property 3 (Pavlov et. al., 2005). *If system (9.6) is exponentially convergent, then, for any input $\boldsymbol{k} \in \overline{\mathbb{PC}}_d$, $\boldsymbol{K} \in \mathbb{M}_d$, the steady-state solution $\bar{\boldsymbol{x}}_k(t)$ is the only solution defined and bounded for almost all $t \in \mathbb{R}$.*

We note that this property was formulated in [Pavlov *et. al.*, 2005] for differential equations.

9.5 Convergence properties of Lur'e-type measure differential inclusions

In this section we study the convergence properties for perturbed Lur'e-type measure differential inclusions of the following form:

$$\mathrm{d}\boldsymbol{x} = \boldsymbol{A}\boldsymbol{x}\,\mathrm{d}t + \boldsymbol{B}\,\mathrm{d}\boldsymbol{w}(t) + \boldsymbol{D}\,\mathrm{d}\boldsymbol{s},$$
$$-\mathrm{d}\boldsymbol{s} \in \mathcal{H}(\boldsymbol{y})\,\mathrm{d}t + \mathcal{H}(\boldsymbol{y}^+)\,\mathrm{d}\eta, \qquad (9.7)$$
$$\boldsymbol{y} = \boldsymbol{C}\boldsymbol{x},$$

with $\boldsymbol{A} \in \mathbb{R}^{n \times n}$, $\boldsymbol{B} \in \mathbb{R}^{n \times d}$, $\boldsymbol{C} \in \mathbb{R}^{m \times n}$, $\boldsymbol{D} \in \mathbb{R}^{n \times m}$ and $\boldsymbol{x} \in \mathbb{R}^n$ is the system state. Moreover, $\mathrm{d}\boldsymbol{s} = \boldsymbol{\lambda}\,\mathrm{d}t + \boldsymbol{\Lambda}\,\mathrm{d}\eta$ and $\mathcal{H}(\boldsymbol{y})\,\mathrm{d}t + \mathcal{H}(\boldsymbol{y}^+)\,\mathrm{d}\eta$ is the differential measure of the nonlinearity in the feedback loop that is characterised by the set-valued maximal monotone mapping $\mathcal{H}(\boldsymbol{y})$ with $\boldsymbol{0} \in \mathcal{H}(\boldsymbol{0})$. These properties of $\mathcal{H}(\boldsymbol{y})$ imply that $\boldsymbol{y}^\mathrm{T}\boldsymbol{h} \geq \boldsymbol{0}$ for all $\boldsymbol{h} \in \mathcal{H}(\boldsymbol{y})$ and

$\boldsymbol{y} \in \{\boldsymbol{y} \in \mathbb{R}^m | \boldsymbol{y} = \boldsymbol{Cx} \wedge \boldsymbol{x} \in \mathcal{X}\}$, i.e. the action of $\mathcal{H}$ is passive. Furthermore, the inclusion in (9.7) indicates that $\boldsymbol{\lambda} \in -\mathcal{H}(\boldsymbol{y})$ and $\boldsymbol{\Lambda} \in -\mathcal{H}(\boldsymbol{y}^+)$. Finally, the differential measure of the time-dependent perturbation is decomposed as $\mathrm{d}\boldsymbol{w}(t) = \boldsymbol{k}(t)\,\mathrm{d}t + \boldsymbol{K}(t)\,\mathrm{d}\eta$, where $\boldsymbol{k}(t) \in \overline{\mathbb{PC}}_d$, $\boldsymbol{K}(t) \in \mathbb{M}_d$ are functions that represent the non-impulsive and impulsive parts of the perturbation, respectively.

In the remainder of this work we assume that the Lur'e-type measure differential inclusion is consistent as formalised in the following assumption.

Assumption 9.1. The measure differential inclusion (9.7) is consistent.

In the following theorem, we state conditions under which system (9.7) is exponentially convergent. Later, we will exploit this property to solve the tracking control problem. However, since the convergence property has been shown to be beneficial in a wider context, for example in the scope of output regulation, observer design and performance analysis for nonlinear systems, we state this result separately here.

Theorem 9.1. *Consider a measure differential inclusion of the form (9.7), which satisfies Assumption 9.1, with $\mathcal{H}(\boldsymbol{y})$ a (set-valued) maximal monotone mapping with $\boldsymbol{0} \in \mathcal{H}(\boldsymbol{0})$. If the following conditions are satisfied:*

(1) the triple $(\boldsymbol{A}, \boldsymbol{D}, \boldsymbol{C})$ is strictly passive. In other words, there exists a positive definite matrix $\boldsymbol{P} = \boldsymbol{P}^{\mathrm{T}} > 0$ and $\alpha > 0$ for which the following conditions are satisfied:

$$\boldsymbol{A}^{\mathrm{T}}\boldsymbol{P} + \boldsymbol{P}\boldsymbol{A} \leq -2\alpha\boldsymbol{P}, \ \boldsymbol{D}^{\mathrm{T}}\boldsymbol{P} = \boldsymbol{C}. \tag{9.8}$$

(2) there exists a $\beta \in \mathbb{R}$ such that $(\boldsymbol{x}^+)^{\mathrm{T}}\boldsymbol{P}\boldsymbol{B}\boldsymbol{K}(t) \leq \beta$ for all $\boldsymbol{x} \in \mathcal{X}$ and $\boldsymbol{P}$ satisfying (9.8); i.e. the energy input of the impulsive inputs is bounded from above,

(3) the time instances t_i for which the input is impulsive, i.e. for which $\boldsymbol{K}(t)$ is non-zero, are separated by the dwell-time $\tau \leq t_{i+1} - t_i$, with

$$\tau = \frac{\delta}{2(\delta - 1)\alpha} \ln(1 + \frac{2\beta}{\delta^2\gamma^2}),$$

$$\gamma := \sup_{t \in \mathbb{R}, \boldsymbol{\lambda}(0) \in -\mathcal{H}(\boldsymbol{0})} \left\{ \frac{\|\boldsymbol{B}\boldsymbol{k}(t) + \boldsymbol{D}\boldsymbol{\lambda}(0)\|_P}{\alpha} \right\}, \tag{9.9}$$

for some $\delta > 1$.

then system (9.7), with inputs $\boldsymbol{k}(t)$ and $\boldsymbol{K}(t)$, is exponentially convergent.

Proof.   In this proof, we will show that system (9.7) is exponentially convergent. Hereto, we first show that the all solutions of the system converge to each other exponentially. The next step in the proof of exponential convergence is to show that there exists a unique (steady-state) solution that is bounded on $t \in \mathbb{R}$.

Consider two solutions $x_1(t)$ and $x_2(t)$ of the closed-loop system (9.7) and a Lyapunov candidate function $V = \frac{1}{2}\|x_2 - x_1\|_P^2$, where we adopt the notation $\|\xi\|_P^2 = \xi^{\mathrm{T}} P \xi$. Consequently, the differential measure of V satisfies: $\mathrm{d}V = \frac{1}{2}(x_2^+ + x_2^- - x_1^+ - x_1^-)^{\mathrm{T}} P (\mathrm{d}x_2 - \mathrm{d}x_1)$, with $\mathrm{d}x_i = A x_i \, \mathrm{d}t + D \, \mathrm{d}s_i + B \, \mathrm{d}w(t)$, $i = 1, 2$, where $\mathrm{d}s_i = \lambda_i \, \mathrm{d}t + \Lambda_i \, \mathrm{d}\eta$, with $\lambda_i \in -\mathcal{H}(C x_i)$, $\Lambda_i \in -\mathcal{H}(C x_i^+)$, $i = 1, 2$. The differential measure of V has a density $\dot{V}$ with respect to the Lebesgue measure $\mathrm{d}t$ and a density $V^+ - V^-$ with respect to the atomic differential measure $\mathrm{d}\eta$, i.e.

$$\mathrm{d}V = \dot{V} \, \mathrm{d}t + (V^+ - V^-) \, \mathrm{d}\eta. \tag{9.10}$$

We first evaluate the density $\dot{V}$:

$$\begin{aligned}
\dot{V} &= (x_2 - x_1)^{\mathrm{T}} P(D\lambda_2 + A x_2 - (D\lambda_1 + A x_1)) \\
&= \frac{1}{2}(x_2 - x_1)^{\mathrm{T}} \left((PA + A^{\mathrm{T}} P)(x_2 - x_1) + 2C^{\mathrm{T}}(\lambda_2 - \lambda_1) \right), \quad (9.11) \\
&\leq -\alpha \|x_2 - x_1\|_P^2,
\end{aligned}$$

where we used, firstly, that both solutions x_1 and x_2 correspond to the same perturbation $\mathrm{d}w(t)$, secondly, the fact that (9.8) is satisfied and, thirdly, the fact that the mapping $\mathcal{H}(y)$ is monotone. Subsequently, we consider the jump $V^+ - V^-$ of V: $V^+ - V^- = \frac{1}{2}(x_2^+ + x_2^- - x_1^+ - x_1^-)^{\mathrm{T}} P \left(x_2^+ - x_2^- - x_1^+ + x_1^- \right)$, with $x_i^+ - x_i^- = D\Lambda_i + BK(t)$, $\Lambda_i \in -\mathcal{H}(C x_i^+)$, $i = 1, 2$. Elimination of x_1^- and x_2^- and exploiting the monotonicity of $\mathcal{H}(y)$ gives

$$\begin{aligned}
V^+ - V^- &= (x_2^+ - x_1^+ - \frac{D}{2}(\Lambda_2 - \Lambda_1))^{\mathrm{T}} PD (\Lambda_2 - \Lambda_1) \\
&= (y_2^+ - y_1^+)^{\mathrm{T}} (\Lambda_2 - \Lambda_1) - \frac{1}{2} \|(D\Lambda_2 - D\Lambda_1)\|_P^2 \leq 0,
\end{aligned} \tag{9.12}$$

where we used the matrix equality in (9.8). Using (9.10), (9.11) and (9.12), the differential measure of V satisfies $\mathrm{d}V \leq -2\alpha V \, \mathrm{d}t$, along solutions of (9.7). It therefore holds that V strictly decreases (exponentially) over every Lebesgue non-negligible time-interval as long as $x_2 \neq x_1$. In turn, this implies that all solutions of (9.7) converge to each other exponentially (i.e. the system is exponentially incrementally stable):

$$\|x_2^+(t) - x_1^+(t)\| \leq \sqrt{\frac{\lambda_{\max}(P)}{\lambda_{\min}(P)}} e^{-\alpha(t - t_0)} \|x_2^-(t_0) - x_1^-(t_0)\|, \tag{9.13}$$

for almost all $t \geq t_0$ and where $\lambda_{\max}(P)$ and $\lambda_{\min}(P)$ represent the maximum and minimum eigenvalue of P, respectively.

Let us now show that there exists a unique (steady-state) solution that is bounded on $t \in \mathbb{R}$. Consider, hereto, the Lyapunov candidate function $W = \frac{1}{2}x^\mathrm{T} P x$. The differential measure of W can be decomposed as $\mathrm{d}W = \dot{W}\,\mathrm{d}t + (W^+ - W^-)\,\mathrm{d}\eta$. We first evaluate the density $\dot{W}$:

$$\dot{W} = x^\mathrm{T} P(D\lambda + Ax + Bk(t))$$
$$= x^\mathrm{T} PD(\lambda - \lambda(0)) + \frac{1}{2}x^\mathrm{T}(PA + A^\mathrm{T}P)x \qquad (9.14)$$
$$+ x^\mathrm{T} PBk(t) + x^\mathrm{T} PD\lambda(0),$$

with $\lambda \in \mathcal{H}(y)$ and $\lambda(0) \in \mathcal{H}(0)$. Due to the satisfaction of (9.8) and the monotonicity of $\mathcal{H}(y)$, we have that

$$\dot{W} \leq -\alpha\|x\|_P^2 + \|x\|_P\|Bk(t) + D\lambda(0)\|_P. \qquad (9.15)$$

Note that $\dot{W} < 0$ for x satisfying $\|x\|_P > \gamma$ with γ defined in (9.9). Let us use the fact that the function $-(1 - \frac{1}{\delta})\alpha\|x\|_P^2 > -\alpha\|x\|_P^2 + \gamma\alpha\|x\|_P$ for $\|x\|_P > \delta\gamma$, where $\delta > 1$ is an arbitrary constant and $\gamma > 0$. It therefore holds that

$$\dot{W} \leq -2\left(1 - \frac{1}{\delta}\right)\alpha W \quad \text{for } \|x\|_P \geq \delta\gamma,\ \delta > 1. \qquad (9.16)$$

Subsequently, we consider the jump $W^+ - W^-$ of W: $W^+ - W^- = \frac{1}{2}(x^+ + x^-)^\mathrm{T}P(x^+ - x^-)$, with $x^+ - x^- = D\Lambda + BK(t)$ and $\Lambda \in -\mathcal{H}(Cx^+)$. Elimination of x^-, exploiting the passivity of $\mathcal{H}(y)$ and using the matrix equality in (9.8) gives

$$W^+ - W^- = \frac{1}{2}(2x^+ - D\Lambda - BK(t))^\mathrm{T}P(D\Lambda + BK(t))$$
$$= (x^+)^\mathrm{T}(PD\Lambda + PBK(t)) - \frac{1}{2}\|D\Lambda + BK(t)\|_P^2 \qquad (9.17)$$
$$\leq y^{+\mathrm{T}}\Lambda + x^{+\mathrm{T}}PBK(t) \leq \beta,$$

in which we used condition 2 in the theorem. Then, due to (9.16), for the non-impulsive part of the motion it holds that if $\|x(t_0)\|_P \leq \delta\gamma$ then $\|x(t)\|_P \leq \delta\gamma$ for all $t \in [t_0, t^*]$ (if no state resets occur in this time interval). Moreover, as far as the state resets are concerned, (9.17) shows that a state reset from a state $x^-(t_i) \in \mathcal{V}$ with $\mathcal{V} = \{x \in \mathcal{X} \mid \|x\|_P \leq \delta\gamma\}$ can only occur to $x^+(t_i)$ such that $W(x^+(t_i)) := \frac{1}{2}\|x^+(t_i)\|_P^2 \leq W(x^-(t_i)) + \beta \leq \frac{1}{2}\delta^2\gamma^2 + \beta$. During the following open time-interval (t_i, t_{i+1}) for which $K(t) = 0$, the function W evolves as $W(x^-(t_{i+1})) = W(x^+(t_i)) + \int_{(t_i, t_{i+1})} \mathrm{d}W$, which may involve impulsive motion due to dissipative impulses $\mathbf{\Lambda}$. Let $t_\mathcal{V} \in (t_i, t_{i+1})$ be the time-instance for which $\|\mathbf{x}^-(t_\mathcal{V})\|_P = \delta\gamma$. The function W will necessarily decrease during the time-interval $(t_i, t_\mathcal{V})$ due to (9.16) and $W^+ - W^- = (\mathbf{x}^+)^{\mathrm{T}}(\mathbf{PD\Lambda}) - \frac{1}{2}\|\mathbf{D\Lambda}\|_P^2 \leq 0$ (the state-dependent impulses are passive due to passivity of $\mathcal{H}(\mathbf{y})$). It therefore holds that

$$W(\mathbf{x}^-(t_\mathcal{V})) \leq e^{-2(1-\frac{1}{\delta})\alpha(t_\mathcal{V}-t_i)}W(\mathbf{x}^+(t_i)), \tag{9.18}$$

because $\mathrm{d}W \leq -2(1 - \frac{1}{\delta})\alpha W\,\mathrm{d}t + (W^+ - W^-)\,\mathrm{d}\eta \leq -2(1 - \frac{1}{\delta})\alpha W\,\mathrm{d}t$ for positive measures. Using $W(\mathbf{x}^-(t_\mathcal{V})) = \frac{1}{2}\delta^2\gamma^2$ and $W(\mathbf{x}^+(t_i)) \leq \frac{1}{2}\delta^2\gamma^2 + \beta$ in the exponential decrease (9.18) gives $\frac{1}{2}\delta^2\gamma^2 \leq e^{-2(1-\frac{1}{\delta})\alpha(t_\mathcal{V}-t_i)}(\frac{1}{2}\delta^2\gamma^2 + \beta)$ or $t_\mathcal{V}-t_i \leq \frac{\delta}{2(\delta-1)\alpha}\ln(1+\frac{2\beta}{\delta^2\gamma^2})$. Consequently, if the next impulsive time-instance t_{i+1} of the input is after $t_\mathcal{V}$, then the solution $\mathbf{x}(t)$ has enough time to reach $\mathcal{V}$. Hence, if the impulsive time-instance of the input are separated by the dwell-time τ given in (9.9), i.e. $t_{i+1} - t_i \geq \tau$, then the set

$$\mathcal{W} = \left\{ \mathbf{x} \in \mathcal{X} \mid \frac{1}{2}\|\mathbf{x}\|_P^2 \leq \frac{1}{2}\delta^2\gamma^2 + \beta \right\} \tag{9.19}$$

is a compact positively invariant set. Since the size of this positively invariant set is of no concern we can take the limit of the expression for τ in (9.9) for $\delta \to \infty$: $\lim_{\delta\to\infty} \frac{\delta}{2(\delta-1)\alpha}\ln(1 + \frac{2\beta}{\delta^2\gamma^2}) = 0$, which indicates that the dwell-time can be taken arbitrarily small. It therefore suffices to assume that the impulsive inputs $\mathbf{K}(t)$ are separated in time (as required in the theorem) to conclude that the system exhibits a compact positively invariant set, defined in (9.19).

Now, we use Lemma 2 in [Yakubovich, 1964], which formulates that if a dynamic system exhibits a compact positively invariant set, then the existence of a solution that is bounded for $t \in \mathbb{R}$ is guaranteed. We will denote this 'steady-state' solution by $\bar{\mathbf{x}}(t)$. The original lemma is formulated for differential equations (possibly with discontinuities, therewith including differential inclusions, with bounded right-hand sides). Here, we use this lemma for measure differential inclusions and would like to note that the proof of the lemma allows for such extensions if we only require continuous dependence on initial conditions. The latter is guaranteed for the class of Lur'e-type measure differential inclusions under study, because the system is exponentially incrementally stable (as shown in the first part of the proof), which implies continuous dependence on initial conditions.

The combination of the fact that all solutions are exponentially stable with the fact that there exists a (steady-state) solution $\bar{\mathbf{x}}(t)$, of locally bounded variation, which is bounded for $t \in \mathbb{R}$ for which it is defined, completes the proof of the fact that the system (9.7) is exponentially convergent according to Definition 9.5. $\qquad\square$

9.6   Tracking control of Lur'e-type measure differential inclusions

In this section we study the tracking control problem for Lur'e-type measure differential inclusions of the following form:

$$
\begin{aligned}
\mathrm{d}\boldsymbol{x} &= \boldsymbol{A}_{ol}\boldsymbol{x}\,\mathrm{d}t + \boldsymbol{B}\,\mathrm{d}\boldsymbol{u} + \boldsymbol{D}\,\mathrm{d}\boldsymbol{s}, \\
-\mathrm{d}\boldsymbol{s} &\in \boldsymbol{\mathcal{H}}(\boldsymbol{y})\,\mathrm{d}t + \boldsymbol{\mathcal{H}}(\boldsymbol{y}^{+})\,\mathrm{d}\eta, \qquad (9.20) \\
\boldsymbol{y} &= \boldsymbol{C}\boldsymbol{x},
\end{aligned}
$$

with $\boldsymbol{A}_{ol} \in \mathbb{R}^{n \times n}$ and $\mathrm{d}\boldsymbol{u} = \boldsymbol{p}\,\mathrm{d}t + \boldsymbol{P}\,\mathrm{d}\eta$ is the differential measure of the control action. Herein, $\boldsymbol{p}$ represents the non-impulsive part of the control action and $\boldsymbol{P}$ represents its impulsive part.

The tracking problem considered in this work is formalised as follows:

Tracking problem:
Design a control law for $\mathrm{d}\boldsymbol{u}$ that, based on information on the desired state trajectory $\boldsymbol{x}_d(t)$ and the measured state $\boldsymbol{x}$, renders $\boldsymbol{x}(t) \to \boldsymbol{x}_d(t)$ as $t \to \infty$ and the states of the closed-loop system are bounded.

To solve this problem, we adopt the following assumption:

Assumption 9.2. The desired trajectory $\boldsymbol{x}_d(t)$ is a function of locally bounded variation and there exists $\mathrm{d}\boldsymbol{u}_{ff}(t) = \boldsymbol{p}_{ff}(t)\,\mathrm{d}t + \boldsymbol{P}_{ff}(t)\,\mathrm{d}\eta$, with both $\boldsymbol{p}_{ff}(t) \in \overline{\mathbb{PC}}_d$ and $\boldsymbol{P}_{ff}(t) \in \mathbb{M}_d$, such that $\boldsymbol{x}_d(t)$ satisfies

$$
\begin{aligned}
\mathrm{d}\boldsymbol{x}_d(t) &= \boldsymbol{A}_{ol}\boldsymbol{x}_d(t)\,\mathrm{d}t + \boldsymbol{B}\,\mathrm{d}\boldsymbol{u}_{ff}(t) + \boldsymbol{D}\,\mathrm{d}\boldsymbol{s}, \\
-\mathrm{d}\boldsymbol{s} &\in \mathrm{d}\boldsymbol{\mathcal{H}}(\boldsymbol{C}\boldsymbol{x}_d(t))\,\mathrm{d}t + \boldsymbol{\mathcal{H}}(\boldsymbol{C}\boldsymbol{x}_d^{+}(t))\,\mathrm{d}\eta,
\end{aligned} \qquad (9.21)
$$

i.e. $\mathrm{d}\boldsymbol{u}_{ff}(t)$ can be considered to be a reference control (feedforward) generating $\boldsymbol{x}_d(t)$.

When addressing the tracking problem, it is commonly split in two parts: firstly, finding the appropriate feedforward and, secondly, stabilising the desired solution. In the current paper, we primarily focus on the second problem. Note that also for smooth systems the existence of the feedforward is a natural assumption (think of the solvability of the regulator equations as a natural assumption in the scope of output regulation [Isidori and Byrnes, 1990; Pavlov et al., 2005].

We propose to tackle the tracking problem by means of a combination of Lebesgue measurable linear error-feedback and a possibly impulsive feedforward control:

$$
\mathrm{d}\boldsymbol{u} = \boldsymbol{u}_{fb}(\boldsymbol{x}, \boldsymbol{x}_d(t))\,\mathrm{d}t + \mathrm{d}\boldsymbol{u}_{ff}(t), \qquad (9.22)
$$

with

$$u_{fb}(x, x_d(t)) = N (x - x_d(t)),$$
$$du_{ff}(t) = p_{ff}(t) \, dt + P_{ff}(t) \, d\eta, \tag{9.23}$$

where $N \in \mathbb{R}^{d \times n}$ is the feedback gain matrix. We restrict the energy input of the impulsive control action $P_{ff}(t)$ to be bounded from above: $(x^+)^T B P_{ff} \leq \beta$. Note that this condition puts a bound on the jumps in the desired trajectory $x_d(t)$ which can be realised. Combining the control law (9.22) with the system dynamics (9.20) yields the closed-loop dynamics:

$$dx = Ax \, dt + D \, ds + B(-Nx_d(t) \, dt + du_{ff}(t)),$$
$$-ds \in \mathcal{H}(y) \, dt + \mathcal{H}(y^+) \, d\eta, \tag{9.24}$$
$$y = Cx,$$

with $A = A_{ol} + BN$.

In Theorem 9.2 stated below, the convergence property of the closed-loop system is exploited to solve the tracking problem. The main idea of this convergence-based control design of the form (9.22) is that it guarantees the following two properties of the closed-loop system:

(a) the closed-loop system (9.24) has a trajectory which is bounded for all t and along which the tracking error $x - x_d(t)$ is identically zero. In other words, the feedforward $du_{ff}(t)$ has to be designed such that it induces the desired solution $x_d(t)$;

(b) the closed-loop system (9.24) is exponentially convergent. Hereto, the control gain matrix N should be designed appropriately.

Condition b) guarantees that the closed-loop system has a unique bounded globally exponentially stable steady-state solution, while condition a) guarantees that, by Property 3, this steady-state solution equals the bounded solution of the closed-loop system with zero tracking error.

Theorem 9.2. *Consider a measure differential inclusion of the form (9.20), with $\mathcal{H}(y)$ a (set-valued) maximal monotone mapping with $0 \in \mathcal{H}(0)$. Consider the (impulsive) control design (9.22), (9.23). Suppose the desired trajectory $x_d(t)$ satisfies Assumption 9.2 with $du_{ff}(t)$ as in (9.23) being the corresponding feedforward. If the resulting closed-loop system (9.24) satisfies the conditions in Theorem 9.1, with $k(t) := -Kx_d(t) + p_{ff}(t)$ and $K(t) := P_{ff}(t)$, then the desired solution $x_d(t)$ is a globally exponentially stable solution of the closed-loop system (9.24), i.e. the tracking problem is solved.*

Proof. Since the closed-loop system (9.24) satisfies the conditions of Theorem 9.1, all solutions of the closed-loop system (9.24) converge to each other exponentially, see (9.13) in the proof of Theorem 9.1. Since the desired solution is a solution of (9.24), for $x(0) = x_d(0)$, by the choice of the feedforward, see Assumption 9.2, the desired solution is a globally exponentially stable solution of (9.24). $\qquad\square$

9.7 Example of a mechanical system with a unilateral constraint

Let us consider a mechanical system consisting of two inertias, m_1 and m_2, which are coupled by a linear spring c and a linear damper b_1, see Figure 9.1. The inertia m_1 is attached to the earth by a linear damper b_2 and m_2 is subject to a one-way clutch. Moreover, m_1 is actuated by a (possibly impulsive) control force $\mathrm{d}u$. The open-loop dynamics is described

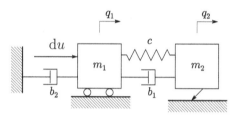

Fig. 9.1   Motor-load configuration with one-way clutch and impulsive actuation.

by (9.20) with

$$
A_{ol} = \begin{bmatrix} 0 & -1 & 1 \\ \frac{c}{m_1} & -\frac{b_1+b_2}{m_1} & \frac{b_1}{m_1} \\ -\frac{c}{m_2} & \frac{b_1}{m_2} & -\frac{b_1}{m_2} \end{bmatrix}, \quad B = \begin{bmatrix} 0 \\ \frac{1}{m_1} \\ 0 \end{bmatrix}, \tag{9.25}
$$

$D^{\mathrm{T}} = \begin{bmatrix} 0 & 0 & \frac{1}{m_2} \end{bmatrix}$ and $C = \begin{bmatrix} 0 & 0 & 1 \end{bmatrix}$. The state vector is given by $x = \begin{bmatrix} q_2 - q_1 & u_1 & u_2 \end{bmatrix}^{\mathrm{T}}$, with q_1 and q_2 the displacements of m_1 and m_2, respectively, and u_1 and u_2 the velocities of m_1 and m_2, respectively. The differential measure $\mathrm{d}s = \mathcal{H}(y)\,\mathrm{d}t + \mathcal{H}(y^+)\,\mathrm{d}\eta$ of the force in the one-way clutch is characterised by the scalar set-valued maximal monotone mapping $\mathcal{H}(x) = \mathrm{Upr}(x)$. The set-valued function $\mathrm{Upr}(x)$ is the unilateral primitive [Glocker, 2001]: $-y \in \mathrm{Upr}(x) \Leftrightarrow 0 \le x \perp y \ge 0 \Leftrightarrow x \ge 0,\ y \ge 0,\ xy = 0$, being a maximal monotone operator. We adopt the following system parameters: $m_1 = m_2 = 1$, $c = 10$, $b_1 = 1$ and $b_2 = -1.4$.

The desired velocity of the second mass is a periodic sawtooth wave with period time T:

$$x_{d3}(t) = \begin{cases} \mathrm{mod}\,(t,T) & \text{for } 0 \le \mathrm{mod}\,(t,T) \le \frac{T}{4} \\ \frac{T}{2} - \mathrm{mod}\,(t,T) & \text{for } \frac{T}{4} \le \mathrm{mod}\,(t,T) \le \frac{T}{2} \\ 0 & \text{for } \frac{T}{2} \le \mathrm{mod}\,(t,T) \le T \end{cases},$$

where these equations represent a ramp-up, ramp-down, and a deadband phase, respectively. The signal $x_{d3}(t)$ for $T = 1$ s is shown by the dotted line in Fig. 9.2. The desired trajectory $x_{d3}(t)$ is a periodic signal which is time-continuous but has three kinks in each period. Kinks in $x_{d3}(t)$ can be achieved by applying an impulsive force on the first mass which causes an instantaneous change in the velocity $x_2 = u_1$ and therefore a discontinuous force in the damper b_1. The one-way clutch on the second mass prevents negative values of x_{d3} and no impulsive force on the first mass is therefore necessary for the change from ramp-down to deadband. In a first step, the signals $x_{d1}(t)$, $x_{d2}(t)$ and $ds(t)$ are designed such that

$$\dot{x}_{d1}(t) = -x_{d2}(t) + x_{d3}(t)$$

$$\mathrm{d}x_{d3}(t) = \left(-\frac{c}{m_2}x_{d1}(t) - \frac{b_1}{m_2}\left(-x_{d2}(t) + x_{d3}(t) \right) \right)\mathrm{d}t$$

$$+ \frac{1}{m_2}\,\mathrm{d}s(t), \tag{9.26}$$

with

$$-\,\mathrm{d}s(t) \in \mathrm{Upr}(x_{d3}(t))\,\mathrm{d}t + \mathrm{Upr}(x_{d3}^{+}(t))\,\mathrm{d}\eta,$$

for the given periodic trajectory $x_{d3}(t)$. The solution of this problem is not unique as we are free to chose $ds(t) \ge 0$ for $x_{d3}(t) = 0$. By fixing $ds(t) = \dot{s}_0\,\mathrm{d}t$ to a constant value for $x_{d3}(t) = 0$ (i.e. $\dot{s}_0$ is a constant), we obtain the following discontinuous differential equation for $x_{d1}(t)$:

$$\dot{x}_{d1} = \begin{cases} \frac{m_2}{b_1}(-\dot{x}_{d3}(t) - \frac{c}{m_2}x_{d1}) & x_{d3}(t) > 0, \\ \frac{m_2}{b_1}(-\dot{x}_{d3}(t) - \frac{c}{m_2}x_{d1} + \frac{1}{m_2}\dot{s}_0) & x_{d3}(t) = 0. \end{cases} \tag{9.27}$$

The numerical solution of $x_{d1}(t)$ gives (after a transient) a periodic signal $x_{d1}(t)$ and $x_{d2}(t) = -\dot{x}_{d1}(t) + x_{d3}(t)$ (see the dotted lines in Figs. 9.4 and 9.5 which are mostly hidden by the solid lines). We have taken $\dot{s}_0 = 1$. Subsequently, the feedforward input $\mathrm{d}u_{ff} = p_{ff}\,\mathrm{d}t + P_{ff}\,\mathrm{d}\eta$ is designed such that

$$\mathrm{d}u_{ff} = m_1\,\mathrm{d}x_{d2} - \left(cx_{d1} + b_1(-x_{d2} + x_{d3}) - b_2 x_{d2} \right)\mathrm{d}t \tag{9.28}$$

and it therefore holds that $x(t) = x_d(t)$ for $t \geq 0$ if $x(0) = x_d(0)$, where $x(t)$ is a solution of (9.20), (9.25), with $du = du_{ff}$. The feedforward input du_{ff}/dt is shown in Fig. 9.6 and is equal to $p_{ff}(t)$ almost everywhere. Two impulsive inputs $P_{ff}(t)$ per period can be seen at the time-instances for which there is a change from 'ramp-up to ramp-down' and from 'deadband to ramp-up'. Next, we implement the control law (9.22) on system (9.20) with the feedforward du_{ff} as in (9.28). We choose $N = \begin{bmatrix} 0 & -4 & 0 \end{bmatrix}$ which ensures that (9.8) is satisfied with

$$P = \begin{bmatrix} 34 & -10.5 & 0 \\ -10.5 & 6 & 0 \\ 0 & 0 & 1 \end{bmatrix}, \quad \alpha = 0.25. \tag{9.29}$$

Consequently, the closed-loop system (9.20), (9.25), (9.22), (9.28) is exponentially convergent. Fig. 9.2 shows the closed-loop dynamics (in terms of x_3) for which the desired periodic solution $x_d(t)$ is globally exponentially stable. Fig. 9.3 shows the open-loop dynamics (in terms of x_3) for which there is no state-feedback. Without feedback, the desired periodic solution $x_d(t)$ is not globally attractive, not even locally, and the solution from the chosen initial condition is attracted to a stable period-2 solution. Clearly, the system without feedback is not convergent. For both cases the initial condition $x(0) = \begin{bmatrix} 0.16 & 2.17 & 0 \end{bmatrix}^{\mathrm{T}}$ was used. Figs. 9.4 and 9.5 show the time-histories of $x_1(t)$ and $x_{d1}(t)$, respectively $x_2(t)$ and $x_{d2}(t)$, in solid and dotted lines. Jumps in the state $x_2(t)$ and desired state $x_{d2}(t)$ can be seen on time-instances for which the feedforward input is impulsive.

9.8   Conclusions

In this paper, we have studied, firstly, the stability of time-varying solutions of perturbed Lur'e-type measure differential inclusions and, secondly, the tracking control problem for this class of systems. The framework of measure differential inclusions allows us to describe systems with discontinuities in the state evolution, such as mechanical systems with unilateral constraints. In the scope of the tracking problem, the stability properties of time-varying solutions play a central role. Therefore, we have presented results on the stability of time-varying solutions of such systems in the terms of the convergence property. Next, this property is exploited to provide a solution to the tracking problem, where the desired solution may exhibit state jumps. The results are illustrated by application to a mechanical motion system with a unilateral velocity constraint.

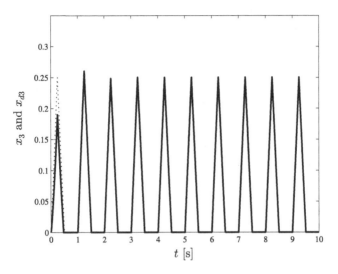

Fig. 9.2 $x_3(t)$ (solid) and $x_{d3}(t)$ (dotted) for the case of feedback and feedforward control.

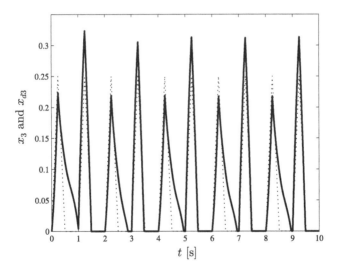

Fig. 9.3 $x_3(t)$ (solid) and $x_{d3}(t)$ (dotted) for the case of only feedforward control.

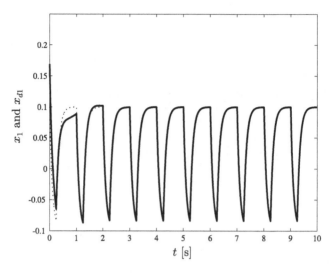

Fig. 9.4 $x_1(t)$ (solid) and $x_{d1}(t)$ (dotted) for the case of feedback and feedforward control.

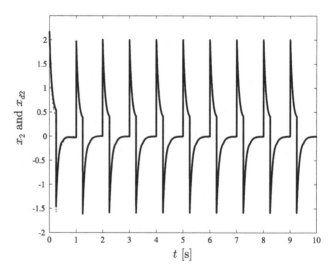

Fig. 9.5 $x_2(t)$ (solid) and $x_{d2}(t)$ (dotted) for the case of feedback and feedforward control.

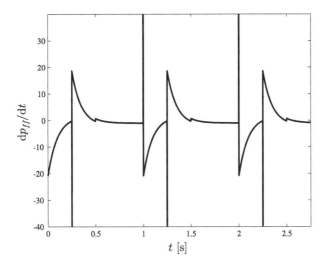

Fig. 9.6   Feedforward $\mathrm{d}u_{ff}/\mathrm{d}t$.

References

Acary V.,Brogliato B., and Goeleven D. (2008), "Higher order Moreaus sweeping process: mathematical formulation and numerical simulation," *Mathematical Programming, Series A*, vol. 113, pp. 133–217.

Angeli D. (2002), "A Lyapunov approach to incremental stability properties." *IEEE Trans. on Automatic Control* **47**, 410–421.

Aubin J.-P. and Frankowska H. (1990), "Set-valued Analysis", vol. 2 of *Systems and Control: Foundations and Applications*. Birkhäuser, Boston.

Bourgeot J. M. and Brogliato B. (2005), "Tracking control of complementarity Lagrangian systems," *Int. J. Bifurcation and Chaos*, vol. 15, no. 6, pp. 1839–1866.

Brogliato B. (1999), *Nonsmooth Mechanics*, 2nd ed.   London: Springer.

Brogliato B. (2004), "Absolute stability and the Lagrange-Dirichlet theorem with monotone multivalued mappings," *Systems & Control Letters*, vol. 51, pp. 343–353.

Brogliato B., Niculescu S.-I., and Orhant P. (1997), "On the control of finite-dimensional mechanical systems with unilateral constraints," *IEEE Trans. on Automatic Control*, vol. 42, no. 2, pp. 200–215.

Brogliato B., and Heemels W.P.M.H. (2009), "Observer design for Lure systems with multivalued mappings: a passivity approach", *IEEE Transactions on Automatic Control*, Accepted.

Camlibel K., van de Wouw N. (2007), "On the convergence of linear passive complementarity systems", *in the Proceedings of the 46th IEEE Conference* on Decision and Control (CDC2007), New Orleans, U.S.A., pp. 5886-5891.

Demidovich B. P. (1967), Lectures on Stability Theory (in Russian). Moscow: Nauka.

Filippov A.F. (1988), Differential Equations with Discontinuous Right-hand Sides. Kluwer Academic, Dordrecht.

Fromion V., Monaco S., and Normand-Cyrot D., (1996) "Asymptotic properties of incrementally stable systems." IEEE Trans. on Automatic Control 41, 721–723.

Glocker Ch. (2001), Set-Valued Force Laws, Dynamics of Non-Smooth Systems, ser. Lecture Notes in Applied Mechanics. Berlin: Springer-Verlag, vol. 1.

Glocker Ch. (2005), "Models of non-smooth switches in electrical systems." Int. J. Circuit Theory and Applications **33**, 205–234.

Goebel R., Sanfelice R.G. and Teel A. (2009), "Hybrid Dynamical Systems," IEEE Control Systems Magazine, vol. 29, no. 2, pp. pp. 28–93.

Isidori A. and Byrnes C. I. (1990), "Output regulation of nonlinear systems," IEEE Trans. on Automatic Control, vol. 35, pp. 131–140.

Leine R. I. and Nijmeijer H. (2004), Dynamics and Bifurcations of Non-Smooth Mechanical Systems, vol. 18 of Lecture Notes in Applied and Computational Mechanics. Springer Verlag, Berlin.

Leine R. I. and N. van de Wouw (2008), Stability and Convergence of Mechanical Systems with Unilateral Constraints, ser. Lecture Notes in Applied and Computational Mechanics. Berlin: Springer Verlag, vol. 36.

Lohmiller W., and Slotine J.-J. E. (1998), "On contraction analysis for nonlinear systems." Automatica **34**, 683–696.

Menini L. and Tornambè A. (2001), "Asymptotic tracking of periodic trajectories for a simple mechanical system subject to nonsmooth impacts," IEEE Trans. on Automatic Control, vol. 46, no. 7, pp. 1122–1126.

Monteiro Marques M. (1993), Differential Inclusions in Nonsmooth Mechanical Systems. Basel: Birkhaüser.

Moreau J. J. (1988), "Unilateral contact and dry friction in finite freedom dynamics," in Non-Smooth Mechanics and Applications, ser. CISM Courses and Lectures, J. J. Moreau and P. D. Panagiotopoulos, Eds. Wien: Springer, vol. 302, pp. 1–82.

Moreau J. J. (1988), "Bounded variation in time," in Topics in Nonsmooth Mechanics, J. J. Moreau, P. D. Panagiotopoulos, and G. Strang, Eds. Basel, Boston, Berlin: Birkhäuser Verlag, pp. 1–74.

Pavlov A.V., van de Wouw N., and Nijmeijer H. (2005), Uniform Output Regulation of Nonlinear Systems: A Convergent Dynamics Approach. Boston: Birkhäuser, in Systems & Control: Foundations and Applications (SC) Series.

Pavlov A.V., van de Wouw N., Pogromski A.Y. and Nijmeijer H. (2004) , "Convergent dynamics, a tribute to B.P. Demidovich," Systems and Control Letters, vol. 52, no. 3-4, pp. 257–261.

Pavlov A.V., Pogromsky A.Y., van de Wouw N., Nijmeijer H. (2007), "On convergence properties of piecewise affine systems", International Journal of Control, Volume 80, Issue 8, pages 1233 - 1247.

Pavlov A.V., van de Wouw N. (2008), "Convergent discrete-time nonlinear systems: the case of PWA systems", *in the Proceedings of the 2008 ACC Conference, Seattle, U.S.A.*, pp. 3452-3457.

Pogromsky A. (1998), "Passivity based design of synchronizing systems." *Int. J. Bifurcation and Chaos* **8**, 295–319.

Sontag E. D. (1995),"On the input-to-state stability property." *European J. Control* **1**, 24–36.

van der Schaft A. J. and Schumacher J. M. (2000), *An Introduction to Hybrid Dynamical Systems*, ser. Lecture Notes in Control and Information Sciences. London: Springer, vol. 251.

van de Wouw N. and Pavlov A.V (2008), "Tracking and Synchronisation for a Class of PWA System", *Automatica* Vol. 44, No. 11, pp. 2909-2915.

van de Wouw N., Pastink H.A., Heertjes M.F., Pavlov A.V. and Nijmeijer H. (2008), "Performance of Convergence-based Variable-gain Control of Optical Storage Drives", *Automatica* Vol. 44, No. 1, pp. 15-27.

Willems J. L. (1970), *Stability theory of Dynamical Systems.* London: Thomas Nelson and Sons Ltd..

Yakubovich V. (1964), "Matrix inequalities method in stability theory for nonlinear control systems: I. absolute stability of forced vibrations," *Automation and Remote Control*, vol. 7, pp. 905–917.

Synchronization between Coupled Oscillators: An Experimental Approach

D.J. Rijlaarsdam, A.Y. Pogromsky, H. Nijmeijer

Department of Mechanical Engineering
Eindhoven University of Technology
The Netherlands

Abstract

An experimental set-up is presented that allows to study both controlled and uncontrolled synchronization between a variety of different oscillators. Two experiments are discussed where uncontrolled synchronization between two types of identical oscillators is investigated. First, uncontrolled synchronization between two Duffing oscillators is investigated and second, uncontrolled synchronization between two coupled rotating elements is discussed. In addition to experimental results, analytical and numerical results are presented that support the experimental results.

10.1 Introduction

In the 17th century the Dutch scientist Christiaan Huygens observed a peculiar phenomenon when two pendula clocks, mounted on a common frame, seemed to 'sympathize' as he described it. What he observed was that both clocks adjusted their rhythm towards anti-phase synchronized motion. This effect is now known as frequency or Huygens synchronization and is caused by weak interaction between the clocks due to small displacements of the connecting frame. In [Bennett *et al.*, 2002; Pantale-

one, 2002; Senator, 2006; Kuznetsov *et al.*, 2007] an extended analysis of
this phenomenon is presented. In [Oud *et al.*, 2006] the authors present an
experimental study of Huygens synchronization and finally, in [Pogromsky
et al., 2003; Pogromsky *et al.*, 2006] a study of the uncontrolled as well as
the controlled Huygens experiment is presented.

In this paper an experimental set-up [Tillaart, 2006] is presented that
allows to study both controlled and uncontrolled synchronization between
a variety of different oscillators. In section 10.2 the set-up is introduced
and the dynamical properties of the system are defined. Furthermore, a
way to modify these properties to represent a variety of different oscillators
is presented. Next, section 10.3 discusses an experiment with two synchronizing Duffing oscillators. The stability of the synchronization manifold in
investigated followed by numerical and experimental results. Section 10.4
presents an experiment where the set-up is adjusted to model two rotating
eccentric discs coupled through a third disc mounted on a common axis.
Conclusions and future research are presented in section 10.5.

10.2   Experimental set-up

In order to experimentally study synchronization between coupled oscillators a set-up consisting of two oscillators, mounted on a common frame has
been developed (see figure 10.1 and 10.2).

Fig. 10.1   Photograph of the set-up.

The parameters of primary interest are presented in the table below.

	Oscillator 1	Oscillator 2	Frame / beam
Mass	m	m	M
Stiffness	$\kappa_1(\cdot)$	$\kappa_2(\cdot)$	$\kappa_3(\cdot)$
Damping	$\beta_1(\cdot)$	$\beta_2(\cdot)$	$\beta_3(\cdot)$

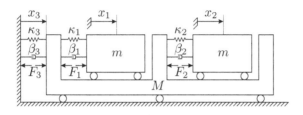

Fig. 10.2   Schematic representation of the set-up.

The set-up contains tree actuators and position sensors on all degrees of freedom. Furthermore, although the masses of the oscillators (m) are fixed, the mass of the connecting beam (M) may be varied by a factor 10. This allows for mechanical adjustment of the coupling strength. A schematic representation of the set-up is depicted in figure 10.2 and the equations of motion are given by:

$$m\ddot{x}_1 = -\kappa_1(x_1 - x_3) - \beta_1(\dot{x}_1 - \dot{x}_3) + F_1 \tag{10.1}$$

$$m\ddot{x}_2 = -\kappa_2(x_2 - x_3) - \beta_2(\dot{x}_2 - \dot{x}_3) + F_2 \tag{10.2}$$

$$M\ddot{x}_3 = \kappa_1(x_1 - x_3) + \kappa_2(x_2 - x_3) \tag{10.3}$$
$$- \kappa_3(x_3) + \beta_1(\dot{x}_1 - \dot{x}_3) + \beta_2(\dot{x}_2 - \dot{x}_3) - \beta_3(\dot{x}_3)$$
$$+ F_3 - F_1 - F_2$$

Here $m, M \in \mathbb{R}_{>0}$ and $x_i \in D_i \subset \mathbb{R}$, $i = 1, 2, 3$ are the masses and displacements of the oscillators and the beam respectively. Functions $\kappa_i : \mathbb{R} \mapsto \mathbb{R}$, $\beta_i : \mathbb{R} \mapsto \mathbb{R}$ describe the stiffness and damping characteristics present in the system. F_i are the actuator forces that may be determined such that the experimental set-up models a large variety of different dynamical systems (see section 10.2.1).

The stiffness and damping in the system are found to be very well approximated by:

$$\kappa_i(q_i) = \sum_{j=1}^{5} k_{ij} q^j \qquad (10.4)$$

$$\beta_i(\dot{q}_i) = b_i \dot{q}_i, \qquad (10.5)$$

where $q_1 = x_1 - x_3$, $q_2 = x_2 - x_3$ and $q_3 = x_3$. The values of k_{ij} and $b_i \ \forall \ i = 1, 2, 3$ have been experimentally obtained and will be used to modify the systems' properties in the sequel.

10.2.1   *Adjustment of the systems' properties*

In order to experiment with different types of oscillators, the derived properties (stiffness and damping) are adjusted. Note that, sinc the damping and stiffness of the system are known and the state of the system is fully measurable, these properties may be adjusted, using actuators, to represent any required dynamics. This allows modeling of different types of springs (linear, cubic), gravity (pendula) and any other desired effect within the limits of the hardware. In the next part of this paper we present two examples of this type of modulation. The system is first adapted to analyze synchronization between Duffing oscillators and second to analyze the synchronizing dynamics of two coupled rotating eccentric discs under the influence of gravity.

10.3   Example 1: Coupled Duffing oscillators

In this section experimental results with respect to two synchronizing Duffing oscillators are presented. After introducing the dynamical system analysis of the limiting behaviour of the system is presented. Finally, both numerical and experimental results are presented and discussed.

10.3.1   *Problem statement and analysis*

Consider the system as depicted in figure 10.3, where

$$\frac{\kappa_d(q_i)}{m} = \omega_0^2 q_i + \vartheta q_i^3 \qquad (10.6)$$

where $q_i = x_i - x_3$ and constants ω_0, $\vartheta \in \mathbb{R}_{>0}$.

The system under consideration represents two undriven, undamped Duffing oscillators coupled through a third common mass. The set-up de-

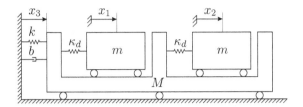

Fig. 10.3 Schematic representation of the set-up modeling two coupled Duffing oscillators.

picted in figure 10.2 can be adjusted to model this system by defining the actuator forces as:

$$F_i = \kappa_i(q_i) + \beta_i(\dot{q}_i) - \kappa_d(q_i), \; i = 1, 2 \tag{10.7}$$

$$F_3 = 0 \tag{10.8}$$

Where $F_3 = 0$ is chosen because, in the original set-up, the beam already models the situation as depicted in figure 10.3 (linear stiffness and damping) fairly accurately. The equations of motion of the resulting system are linear:

$$m\ddot{x}_1 = -\kappa_d(x_1 - x_3) \tag{10.9}$$

$$m\ddot{x}_2 = -\kappa_d(x_2 - x_3) \tag{10.10}$$

$$M\ddot{x}_3 = \kappa_d(x_1 - x_3) + \kappa_d(x_2 - x_3) - kx_3 - b\dot{x}_3, \tag{10.11}$$

where $k, b \in \mathbb{R}_{>0}$ are the stiffness and damping coefficients of the beam.

Before continuing with the experimental and numerical results the system's limiting behaviour is analyzed. In order to do so the notion of anti-phase synchronization needs to be defined. We call the solutions of $x_1(t)$ and $x_2(t)$ asymptotically synchronized in *anti-phase* if they satisfy the following ralation

$$\lim_{t \to \infty} ||x_1(t) - \alpha\sigma_{\left(\frac{T}{2}\right)}x_2(t)|| = 0, \tag{10.12}$$

with α a scale factor and $\sigma_{\left(\frac{T}{2}\right)}$ a shift operator over half an oscillation period (T). Respectively, if instead of the previous relation it follows that

$$\limsup_{t \to \infty} ||x_1(t) - \alpha\sigma_{\left(\frac{T}{2}\right)}x_2(t)|| \leq \varepsilon, \tag{10.13}$$

for some small $\varepsilon > 0$, we say that the solutions are *approximately* asymptotically synchronized in anti-phase. For a more general definition of synchronization the reader can consult [Blekhman *et al.*, 1997].

It can been shown that the dynamics of the oscillators in (10.9)–(10.11) converges to anti-phase synchronization as $t \to \infty$.

To prove this claim consider the system (10.9)–(10.11). To analyze the limit behaviour of this system, the total energy is proposed as a candidate Lyapunov function:

$$\mathcal{V} = \frac{1}{2} \sum_{i=1}^{3} m_i \dot{x}_i^2 + \sum_{i=1}^{3} \int_0^{\xi_i} \kappa_i(s) \, ds, \qquad (10.14)$$

where $m_1 = m_2 = m$, $m_3 = M$, $\xi_i = x_i - x_3$, $i = 1, 2$, $\xi_3 = x_3$, $\kappa_i(q_i) = \kappa_d(q_i)$ and $\kappa_3 = kx_3$. Calculating the time derivative of $\mathcal{V}$ along the solutions of the system (10.9)–(10.11) yields:

$$\dot{\mathcal{V}} = -b\dot{x}_3^2. \qquad (10.15)$$

Hence, we find $\dot{\mathcal{V}} \leq 0$ and the system may be analyzed using LaSalle's invariance principle.

Equation (10.15) implies that $\mathcal{V}$ is a bounded function of time. Moreover, $x_i(t)$ is a bounded function of time and will converge to a limit set where $\dot{\mathcal{V}} = 0$. On this limit set $\dot{x}_3 = \ddot{x}_3 = 0$, according to (10.15). Substituting this in system (10.9)–(10.11) yields $x_3 = 0$ on the system limit set. Substituting $x_3 = \dot{x}_3 = \ddot{x}_3 = 0$ in (10.11) shows:

$$\kappa_d(x_1) = -\kappa_d(x_2) \qquad (10.16)$$

Since κ_d is a one-to-one, odd function, this implies:

$$x_1 = -x_2 \qquad (10.17)$$

Finally, substituting $x_1 = -x_2$ in (10.9)–(10.10) yields:

$$\dot{x}_2 = -\dot{x}_1. \qquad (10.18)$$

Summarizing, it has been shown that any solution of (10.9)–(10.11) will converge to anti-phase synchronized motion according to definition the given definition.

The next paragraph will present numerical and experimental results that support the analysis provided in this section.

10.3.2 *Experimental and numerical results*

In order to experimentally investigate the synchronizing behaviour of two coupled Duffing oscillators the set-up has been modified as specified in the previous section. The oscillators are released from an initial displacement of -3 mm and -2.5 mm respectively (approximately in phase) and allowed to oscillate freely.

Figure 10.4 shows the sum of the positions of the oscillators and the position of the beam versus time. As becomes clear from Figure 10.4, approximate anti-phase synchronization occurs within 40 s. Furthermore, figure 10.5 shows the limiting behaviour of both oscillators and the beam. Although the amplitudes of the oscillators differ significantly, the steady state phase difference is 1.01π. The most probable cause for the amplitude difference is the fact that the oscillators are not exactly identical. As a result, the beam does not come to a complete standstill, although it oscillates with an amplitude that is roughly ten times smaller than that of the oscillators.

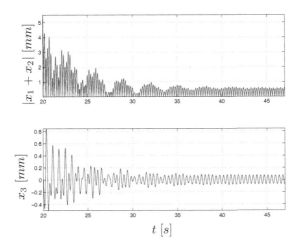

Fig. 10.4   Experimental results: (top) Sum of the displacements of both oscillators. (bottom) Displacement of the connecting beam.

In addition to the experimental results, numerical results are provided in Figure 10.6 and 10.7. The parameters in the simulation are chosen as shown in table 10.2. The results presented in Figure 10.6 and 10.7 correspond to the experimental results provided in 10.4 and 10.5 respectively. Although the oscillation frequencies of the oscillators are almost equal (within 5%) in the simulation and the experiment, the final amplitudes of the oscillators differs by a factor 15. This is due to the fact that in the experiment the damping is over-compensated, resulting in larger amplitudes of the oscillators. In the numerical simulation almost exact anti-phase synchronization with equal oscillator amplitudes is achieved.

Finally, note that some of the differences between the experimental and

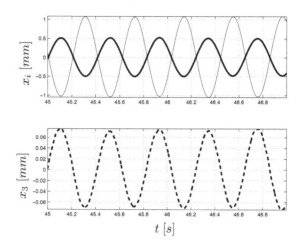

Fig. 10.5 Experimental results: Steady state behaviour of the system. (top) Displacement of the oscillators (- x_1, - x_2). (bottom) Displacement of the connecting beam.

simulation results may be coped with by tuning either the parameters of
the numerical simulation or those of the set-up itself. *The question of
identifying a model can thus be reversed to tuning the parameters of the
set-up rather than those of the model.*

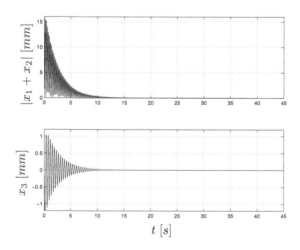

Fig. 10.6 Numerical results: (top) Sum of the displacements of both oscillators. (bottom) Displacement of the connecting beam.

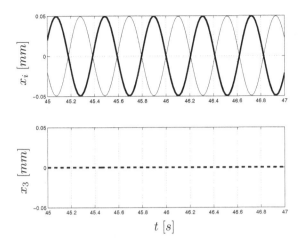

Fig. 10.7   Numerical results: Steady state behaviour of the system. (top) Displacement of the oscillators (- x_1, - x_2). (bottom) Displacement of the connecting beam.

10.4   Example 2: Two coupled rotary disks

Next to the synchronization of Duffing oscillators we investigated synchronization in a system of coupled rotating discs as depicted in Figure 10.8. First the dynamics of the system will be specified in more detail and next experimental results will be presented.

10.4.1   *Problem statement*

Consider the system as depicted in Figure 10.8. This system consists of three discs. Discs 1, 2 represent the oscillators and disc 3 is·connected to both other discs by torsion springs with stiffness k. Each of the discs has an eccentric mass at a distance ℓ_i from it's center ($\ell_1 = \ell_2 = \ell$). Furthermore the middle disc is coupled to the world by a torsion spring with stiffness k_3 and a torsion damper with constant b. The rotation of the discs is represented w.r.t. the world by the angles θ_i.

The equations of motion of the system depicted in Figure 10.8 are:

$$\ddot{\theta}_i = -\vartheta_i \left(k \left(\theta_i - \theta_3 \right) + \delta_i \sin \theta_i \right), \; i = 1, 2 \tag{10.19}$$

$$\ddot{\theta}_3 = \vartheta_3 \Big(\sum_{j=1}^{2} k \left(\theta_j - \theta_3 \right) \tag{10.20}$$

$$- k_3 \theta_3 - b_3 \dot{\theta}_3 - \delta_3 \sin \theta_3 \Big),$$

with $\vartheta_i = \frac{1}{m\ell_i^2 + J_i}$ and $\delta_i = m_i g \ell_i$. The modification to the set-up is now more involved than in the previous example. First of all, the translation coordinates x_i should be mapped to rotation angles θ_i. Secondly, in case of the Duffing oscillator the actuation forces F_1 and F_2 are meant to act on both the oscillators and the connecting mass. In the situation depicted in Figure 10.8 the actuation force generated to model the coupling between the oscillator discs and the middle disc by means of the torsion spring should again act on the oscillators and the connecting beam in our set-up. However, the part of the actuation force that models the influence of gravity on the oscillators should only act on the oscillators and not on the connecting beam, since in Figure 10.8 the gravity on discs 1 and 2 exerts a force only on the corresponding disc and not directly on the middle mass.

In order to adjust the set-up in Figure 10.2 to model the system in figure 10.8 the actuator forces are defined as:

$$F_i = \kappa_i(q_i) + \beta_i(\dot{q}_i) - \vartheta_i\left(\eta_i + g_i\right), i = 1, 2 \qquad (10.21)$$

$$F_3 = \kappa_3(x_3) - \vartheta_3\left(\eta_3 + g_3\right) - \tilde{g}(\cdot), \qquad (10.22)$$

with $\kappa_i(q_i)$ and $\beta_i(\dot{q}_i)$ as defined earlier, $\eta_i = k\left(\theta_i - \theta_3\right)$, $i = 1, 2$, $g_i = \delta_i \sin\theta_i$ and $\tilde{g} = \sum_{j=1}^{2} \vartheta_i g_i$. Damping is left to be the natural damping of the beam in the set-up. Furthermore, translation is mapped to rotation angles according to: $\theta_i = \frac{\pi}{2}\frac{x_i}{x_i^*}$, with x_i^* is the maximal displacement of the oscillators and the beam, assuring $\pm 90°$ turns in the rotation space.

10.4.2 Experimental results

Experimental results, are presented in Figures 10.9 and 10.10. It becomes clear that approximate anti-phase synchronization occurs after about 20 s, like in the Huygens' pendulum set-up. Again complete synchronization does not occur because the oscillators are not identical. In addition figure 10.10 shows the steady state behaviour of the rotating system, from which the approximate anti-phase synchronized behaviour becomes immediately clear.

10.5 Conclusion and future research

An experimental, set-up capable of conducting synchronization experiments with a variety of different oscillators, was presented. Two sets of experi-

$\omega_o = 15.26$	$\vartheta = 8.14$	$M = 0.8$
$m = 1$	$k = 1$	$b = 5$

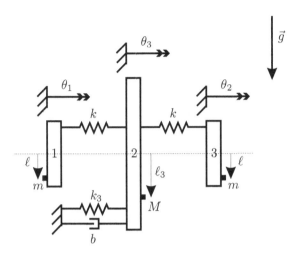

Fig. 10.8 Schematic representation of the set-up modeling two coupled rotating elements.

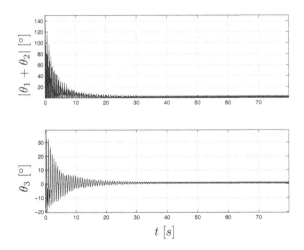

Fig. 10.9 Experimental results: (top) Sum of the rotation angles of the outer discs. (bottom) Rotation angle of the connecting disc.

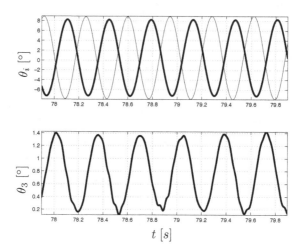

Fig. 10.10   Experimental results: Steady state behaviour of the system. (top) Outer discs (- θ_1, - θ_2). (bottom) Connecting disc.

mental results were provided illustrating the potential of this set-up. First two coupled Duffing oscillators were modeled and synchronizing behaviour was observed in experiments and simulations. Second, it was shown that it is possible to model systems with rotating dynamics and to effectively model the local influence of gravity in this case.

In addition to studying uncontrolled synchronization the set-up has the potential to study controlled synchronization. Furthermore, future work should include modeling the Huygens set-up and perform controlled and uncontrolled synchronization experiments with this type of dynamical system.

Acknowledgments

This work was partially supported by the Dutch-Russian program on interdisciplinary mathematics '*Dynamics and Control of Hybrid Mechanical Systems*' (NWO grant 047.017.018).

References

Bennett, M., M. Schat, H. Rockwood, K. Wiesenfeld (2002). Huygens's clocks, *Proceedings of the Royal Society A: Mathematical, Physical and Engineer-*

ing Sciences **458**, 2019, pp. 563–579. http://dx.doi.org/10.1098/rspa.
2001.0888

Blekhman, I.I., A.L. Fradkov, H. Nijmeijer, A.Y. Pogromsky (1997). On self-synchronization and controlled synchronization, *System & Control Letters* **31**, pp. 299–305.

Kuznetsov, N.V., G.A. Leonov, H. Nijmeijer, A.Y. Pogromsky (2007). Synchronization of Two Metronomes, *3rd IFAC Workshop 'Periodic Control Systems'* (St. Petersburg, Russian Federation).

Oud, W., H. Nijmeijer, A.Yu. Pogromsky (2006). A study of Huygens synchronization. Experimental Results, *Proceedings of the 1st IFAC Conference on Analysis and Control of Chaotic Systems, 2006.*

Pantaleone, J. (2002). Synchronization of metronome, *American Jounal of Physics* **70**, 10, pp. 992–1000.

Pikovsky, A., M. Rosenblum, J. Kurths (2001). *Synchronization, A universal concept in nonlinear sciences*, Cambridge Nonlinear Science Series, (Cambridge University Pres).

Pogromsky A.Y., V.N. Belykh, H. Nijmeijer (2003). Controlled synchronization of pendula, *Proceedings 42rd IEEE Conference on Decision and Control* pp. 4381–4368.

Pogromsky A.Y., V.N. Belykh, H. Nijmeijer (2006). *Group Coordination and Cooperative Control*, Chap. A Study of Controlled Synchronization of Huygens' Pendula. Lecture Notes in Control and Information Sciences (Springer)

Senator, M. (2006). Synchronization of two coupled escapement-driven pendulum clocks, *Journal of Sound and Vibration* **291**, 3–5, pp. 566–603.

Tillaart, M.H.L.M. van den (2006). *Design of a mechanical synchronizing system for research and demonstration purooses for D&C*, Master's thesis, Eindhoven University of Technology, The Netherlands

Swinging Control of Two-pendulum System under Energy Constraints

M. S. Ananyevskiy[*], A.L. Fradkov[*], H. Nijmeijer[†]

[*]Institute for Problems of Mechanical Engineering,
St. Petersburg, Russia

[†]Department of Mechanical Engineering,
Eindhoven University of Technology,
5600 MB, Eindhoven, The Netherlands

Abstract

A method for control of mechanical systems under phase constraints, applicable to energy control of Hamiltonian systems is proposed. The constrained energy control problem for two pendulums by a single control action is studied both analytically and numerically. It is shown that for a proper choice of penalty parameter of the algorithm any energy level for the one pendulum under any specified constraint on the energy of the other pendulum can be achieved. Simulation results confirm fast convergence rate of the algorithm.

11.1 Introduction

Practical control problems for physical and mechanical systems often require taking into account phase constraints. However solving problems with constraints may be very hard, especially for optimal control problems. Constraints which should hold in every time instant may significantly change dynamical nature of the system. Difficulties become still stronger for non-linear and uncertain systems. Complex mechanical systems usually consists of several subsystems. An important practical problem is the selective excitation, when it is needed to increase the energy of one subsystem and to constraint the energy of the other: a passing through resonance [Tomchina *et al.*, 2005], selective molecule excitation [Anan'evskii, 2007], etc.

A unified and powerful method for solving estimation and control problems for nonlinear system is the so called *speed-gradient method* [Fradkov, 1979; Fradkov, 1990; Fradkov and Pogromsky, 1998]. Speed-gradient method applies to the case when the control goal is specified as an asymptotic minimization of a scalar *goal function* of the system state. It allows to design a state feedback allowing to achieve the control goal under certain natural conditions. The method was applied to a variety of nonlinear and adaptive control problems for physical and mechanical systems [Fradkov and Pogromsky, 1998; Fradkov *et al.*, 1999; Fradkov, 2007]. However, taking into account inequality phase constraints were not taken into account before (the case of equality constraints is examined in [Fradkov, 2007]).

In the present paper we propose an approach to control under inequality phase constraints based on an extended version of the speed-gradient method. The constraints are taken into account via a version of the penalty (barrier) functions, well known in the mathematical programming [Fiacco and McCormick, 1990]. A general result providing conditions for achievement of the control goal for constraints specified by a scalar *constraint function* is given. Application of the method is illustrated by example of a pair of pendulums affected by a single controlling force. The goal is to achieve the prespecified value of the one pendulum, while constraint function is the energy of the other pendulum. Both analytical examination and numerical results confirm good performance of the proposed controller.

11.2   Control algorithm: formulation of the problem and approach

Consider a control plant described by a system of states equations

$$\frac{dx}{dt} = F(x, u), \quad x(0) \in \mathbb{X}_0,\ x \in \mathbb{R}^n,\ u \in \mathbb{R}^m,\ t \in \mathbb{R}, \qquad (11.1)$$

here x is the state vector, u is the control input, t is time, $\mathbb{X}_0$ is the set of possible initial states, $F(x, u)$ is smooth function in both arguments. Control goal is specified by means of a non-negative smooth function $Q(x)$

(further it is called *goal function*)

$$\lim_{t \to +\infty} Q(x(t)) = 0, \tag{11.2}$$

where $x(t)$ is the solution of (11.1) with some admissible $u(t)$ and $x(0) \in \mathbb{X}_0$. Phase constraints are specified by an inequality for a smooth function $B(x)$ (further it is called *function of constraints*)

$$B(x(t)) > 0, \quad \text{for all} \quad t \geq 0, \quad \text{and} \quad x(0) \in \mathbb{X}_0. \tag{11.3}$$

It is assumed that $B(x) > 0$ for $x \in \mathbb{X}_0$. The control goal is to minimize goal function $Q(x)$ without crossing set $B^{-1}(0)$[1]. In order to design the control algorithm the idea of speed-gradient algorithm for unconstraint case is used. To this end it is suggested to entroduce the penalty function for minimization without constraints

$$V(x, \alpha) = Q(x) + \frac{\alpha}{B(x)}, \tag{11.4}$$

where $\alpha > 0$ is the penalty parameter. The idea of $V(x, \alpha)$ is similar to penalty (barrier) functions in mathematical programming. If $V(x, \alpha)$ decreases along trajectories of the system (11.1) with some admissible $u(t)$, then according to the property

$$B(x_*) = 0 \Rightarrow \lim_{x \to x_*, B(x) > 0} V(x, \alpha) = +\infty, \tag{11.5}$$

one can conclude that set $B^{-1}(0)$ will be never crossed. Under some additional assumptions (as for interio point method) the minimization of $V(x, \alpha)$ would ensure minimization of $Q(x)$.

According to the speed-gradient method the scalar function $w(x, u, \alpha)$ is introduced

$$w(x, u, \alpha) = \mathcal{L}_F V(x, \alpha) =$$

$$= \left(\frac{\partial Q}{\partial x} - \frac{\alpha}{B(x)^2} \frac{\partial B}{\partial x} \right) F(x, u), \tag{11.6}$$

where $\mathcal{L}_F$ means the derivative along the trajectories of the system (11.1) and $\frac{\partial Q}{\partial x}, \frac{\partial B}{\partial x}$ are the row-vectors of partial derivatives. Then the gradient of $w(x, u, \alpha)$ with respect to input variables is evaluated

$$\nabla_u w(x, u, \alpha) = \left(\frac{\partial w}{\partial u} \right)^T =$$

$$= \left(\frac{\partial F}{\partial u} \right)^T \left(\nabla_x Q(x) - \frac{\alpha}{B(x)^2} \nabla_x B(x) \right). \tag{11.7}$$

[1]$B^{-1}(0) = \{x \in \mathbb{X}_0 : B(x) = 0\}$

Finally, the algoritm of changing $u(t)$ is determined according to the equation

$$u(t) = u_0 - \Gamma \nabla_u w(x(t), u(t), \alpha), \tag{11.8}$$

where u_0 is some initial value of control variable (e.g. $u_0 = 0$), and $\Gamma = \Gamma^T > 0$ is a positive definite gain matrix. If $\alpha = 0$, then (11.8) transformes to the well-known speed-gradient algorithm in finite form without phase constrains. A more general form of (11.8) is

$$u(t) = u_0 - \gamma \psi(x(t), u(t), \alpha), \tag{11.9}$$

where $\gamma > 0$ is a scalar gain parameter and vector-function $\psi(x, u, \alpha)$ satisfies the so-called *pseudogradient* condition

$$\psi(x, u, \alpha)^T \nabla_u w(x, u, \alpha) \geq 0. \tag{11.10}$$

Special case of (11.9) is called *sign-like* or *relay-like* algorithm

$$u(t) = u_0 - \gamma sign \nabla_u w(x(t), u(t), \alpha), \tag{11.11}$$

where sign of a vector is understood component-wise. The solution of differential equation with discontinuous right hand sides is understood in Filippov sense [Filippov, 1988]. Of course, control algorithms (11.8), (11.9), (11.11) are defined only for area where $B(x) > 0$ (because all trajectories starting from $\mathbb{X}_0$ should belong to this area according to phase constraints).

11.2.1   *Special case: Energy control for Hamiltonian systems*

Consider the Hamiltonian system

$$\frac{dp^k}{dt} = -\frac{\partial H(p, q, u)}{\partial q^k}, \quad \frac{dq^k}{dt} = \frac{\partial H(p, q, u)}{\partial p^k},$$

$$k = 1, \ldots, n, \quad (q(0), p(0)) \in \mathbb{X}_0. \tag{11.12}$$

where $q = (q^1, \ldots, q^n)^T$, $p = (p^1, \ldots, p^n)^T$ are the vectors of generalized coordinates and momenta constituting the state vector (p, q) of the system, $H(p, q, u)$ is the controlled Hamiltonian function, $u \in \mathbb{R}^m$ is the input (generalized force), t is time ($t \in \mathbb{R}$), $\mathbb{X}_0$ is the set of possible initial states. Assume that Hamiltonian is affine in control

$$H(p, q) = H_0(p, q) + H_1(p, q)^T u, \tag{11.13}$$

where $H_0(p,q)$ is the internal Hamiltonian and $H_1(p,q)$ is an m−dimensional vector (column) of interaction potentials [Nijmeijer and van der Schaft, 1990]. Complex mechanical systems usually consists of several subsystems. An important practical problem is the selective excitation, when it is needed to increase the energy of one subsystem and to constraint the energy of the other: a passing through resonance [Tomchina *et al.*, 2005], selective molecule excitation [Anan'evskii, 2007], etc. Having in mind these applications split q, p: let q_1, p_1 be vectors of generalized coordinates and momenta of the first subsystem and q_2, p_2 be corresponded vectors of the second one. The following assumption is introduced

$$H_0(p,q) = H_0^1(p_1, q_1) + H_0^2(p_2, q_2) + H_0^{1,2}(p,q), \qquad (11.14)$$

where $H_0^1(p_1, q_1)$ is the Hamiltonian of the first subsystem, $H_0^2(p_2, q_2)$ is the Hamiltonian of the second subsystem, and $H_0^{1,2}(p,q)$ is the Hamiltonian of interaction. Subsystems are called independent, if $H_0^{1,2}(p,q) \equiv 0$.

The selective excitation problem can be formalized as follows. The control goal is to stabilize energy of the first subsystem on the given goal value E_1

$$\lim_{t \to +\infty} H_0^1(p_1(t), q_1(t)) = E_1, \qquad (11.15)$$

the phase constraint is to bound energy of the second subsystem with the given value E_2

$$H_0^2(p_2(t), q_2(t)) < E_2, \quad t \geq 0, \qquad (11.16)$$

where $q_1(t)$, $p_1(t)$, $q_2(t)$, $p_2(t)$ are solutions of (11.12) with some admissible $u(t)$ and $(q(0), p(0)) \in \mathbb{X}_0$.

It is suggested to introduce the goal function and the function of constraints

$$Q(p_1, q_1) = \frac{1}{2} \left(H_0^1(p_1, q_1) - E_1 \right)^2, \qquad (11.17)$$

$$B(p_2, q_2) = E_2 - H_0^2(p_2, q_2). \qquad (11.18)$$

Then the control goal (11.15) takes the form (11.2), and the phase constraint (11.16) takes the form (11.3).

According to (11.4) the penalty function is

$$\begin{aligned} V(p,q,\alpha) &= Q(p_1, q_1) + \frac{\alpha}{B(p_2, q_2)} \\ &= \frac{\left(H_0^1(p_1, q_1) - E_1 \right)^2}{2} + \frac{\alpha}{E_2 - H_0^2(p_2, q_2)}. \end{aligned} \qquad (11.19)$$

For selective excitation problem (11.15), (11.16) of Hamiltonian system (11.12) the control law (11.8) takes form

$$u(t) = u_0 - \Gamma \nabla_u \mathcal{L}_F V(p(t), q(t), \alpha), \qquad (11.20)$$

where $\mathcal{L}_F$ is the derivative along trajectories of the system (11.12), u_0 is some initial value of control variable, and $\Gamma = \Gamma^T > 0$ is a positive definite gain matrix, $\alpha > 0$ is the parameter.

11.3 Two pendulums under a single force

To demonstrate the proposed algorithms an energy selective control problem for two pendulum systems is studed. Consider two independent nonlinear pendulums affected by a single force, schematically depicted in Fig. 11.1. The system equations are as follows

$$\begin{cases} \dot{q}_1 = (ml^2)^{-1} p_1, \\ \dot{p}_1 = -mgl \sin q_1 + ul \cos q_1, \end{cases}$$

$$\begin{cases} \dot{q}_2 = (ml^2)^{-1} p_2, \\ \dot{p}_2 = -mgl \sin q_2 + ul \cos q_2, \end{cases} \qquad (11.21)$$

where dot means the derivative by t; q_1, p_1 are angular coordinate and momentum of the first pendulum, q_2, p_2 are angular coordinate and momentum of the second; g is the gravity acceleration. Pendulums have the same mass m, length l and controlling torque u acts horizontally.

Rewrite system (11.21) in the Hamiltonian form (11.12), (11.13), (11.14). The Hamiltonian function of each pendulum is

$$H_0^k(p_k, q_k) = \frac{1}{2ml^2} p_k^2 + mgl(1 - \cos q_k), \quad k = 1, 2. \qquad (11.22)$$

Pendulums are independent, so

$$H_0^{1,2}(p, q) \equiv 0. \qquad (11.23)$$

The Hamiltonian of interaction potential is

$$H_1(p, q) = -l(\sin q_1 + \sin q_2). \qquad (11.24)$$

The system (11.21) is underactuated, nonlinear and uncontrollable, because if once $q_1(t)$ is equal to $q_2(t)$ and $p_1(t)$ is equal to $p_2(t)$ then they are equal for all t with any control function $u(t)$.

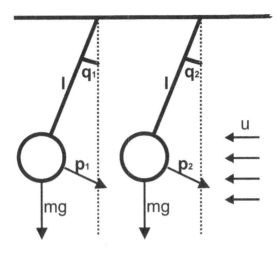

Fig. 11.1   Two independent nonlinear pendulums under a single control force u.

11.3.1   *Control problem formulation*

The problem is to design a controller to swing the first pendulum to the desired energy level E_1 and to constrain the energy of the second one by E_2 during always ($E_2 > 0$). The control objective is formalized by the relation (11.15). The phase constraint is formalized by the relation (11.16). The problem is to find a feedback control law $u(p,q)$ ensuring the control goal (11.15) and phase constraint (11.16).

Suppose that initial conditions of pendulums are different

$$(p_1(0), q_1(0)) \neq (p_2(0), q_2(0)), \qquad (11.25)$$

and satisfy to the phase constraint (11.16)

$$H_0^2(p_2(0), q_2(0)) < E_2. \qquad (11.26)$$

11.3.2   Control algorithm design

According to the approach presented in the previous section the algorithm is designed using the equation (11.20) with $u_0 = 0$

$$u(p, q) = -\Gamma \nabla_u \mathcal{L}_F V(p, q, \alpha)$$

$$= -\Gamma \nabla_u \mathcal{L}_F \left[\frac{\left(H_0^1(p_1, q_1) - E_1\right)^2}{2} + \frac{\alpha}{E_2 - H_0^2(p_2, q_2)} \right]$$

$$= -\Gamma \left(\frac{p_1 \cos q_1}{ml} (H_0^1(p_1, q_1) - E_1) + \alpha \frac{p_2 \cos q_2}{ml} (H_0^2(p_2, q_2) - E_2)^{-2} \right),$$

$$\tag{11.27}$$

where $\Gamma > 0$, $\alpha > 0$ are design parameters.

11.3.3   Control algorithm analysis

Control algorithm analysis is based on Lyapunov function approach and LaSalle principle. Consider $V(p(t), q(t), \alpha)$ as Lyapunov function. The derivative of $V(p, q, \alpha)$ along the system trajectories (11.21) is non-positive

$$\frac{d}{dt} V(p(t), q(t), \alpha) = -\Gamma^{-1} u(p(t), q(t))^2 \le 0. \tag{11.28}$$

Consequently, for any initial conditions from the area $\{(p, q) \colon H_0^2(q_2, p_2) < E_2\}$ the solution of the closed-loop system (11.21), (11.27) is unique and exists[2] for all $t \in [0, +\infty)$. From (11.5) and inequality (11.28) follows, that phase constraint (11.16) always holds.

According to La Salle principle (it is assumed that phase space of each pendulum is cylindrical) any solution tends to the largest invariant set of the closed-loop system (11.21), (11.27). For $(p(t), q(t))$ to be in the invariant set, $\dot{V} = -\Gamma^{-1} u^2 = 0$ which means that

$$\frac{p_1(t) \cos q_1(t)}{ml} (H_0^1(p_1(t), q_1(t)) - E_1)$$

$$+ \alpha \frac{p_2(t) \cos q_2(t)}{ml} (H_0^2(p_2(t), q_2(t)) - E_2)^{-2} \equiv 0 \tag{11.29}$$

---

[2]This note is important, because control function (11.27) is not defined when $H_0^2(p_2, q_2) = E_2$. But according to (11.19), (11.26), (11.28) the trajectories will never cross this set. The right hand of the equations (11.21) is smooth and bounded in the area $\{(p, q) \colon V(p, q, \alpha) \le V(p(0), q(0), \alpha)\}$, so the solution exists and unique for any initial conditions from the area $\{(p, q) \colon H_0^2(p_2, q_2) < E_2\}$ and for all $t \in [0, +\infty)$ (but may be not for $t \in \mathbb{R}$).

must be satisfied with (11.21), (11.27). Since Hamiltonian is constant while $u = 0$, therefore the equation (11.29) can be rewritten

$$\frac{A}{ml^2}p_1(t)\cos q_1(t) + \frac{B}{ml^2}p_2(t)\cos q_2(t) \equiv 0, \qquad (11.30)$$

where the value of A, B

$$A = -l(H_0^1(p_1(t), q_1(t)) - E_1), \qquad (11.31)$$

$$B = -\alpha l(H_0^2(p_2(t), q_2(t)) - E_2)^{-2} \qquad (11.32)$$

are constants. Indeed,

$$\frac{d}{dt}\sin q_k(t) = \frac{1}{ml^2}p_k(t)\cos q_k(t), \quad k = 1, 2, \qquad (11.33)$$

consequenty

$$A\sin q_1(t) + B\sin q_2(t) \equiv const. \qquad (11.34)$$

According to (11.21) for $u = 0$ it follows from (11.34) that

$$A\sin q_1(t) + B\sin q_2(t) \equiv -\frac{A}{mgl}\dot{p}_1(t) - \frac{B}{mgl}\dot{p}_2(t) \equiv const. \quad (11.35)$$

It follows from (11.28) that $p(t)$ is a bounded function for $t \geq 0$. Consequently

$$A\sin q_1(t) + B\sin q_2(t) \equiv -\frac{A}{mgl}\dot{p}_1(t) - \frac{B}{mgl}\dot{p}_2(t) \equiv 0. \quad (11.36)$$

From the properties of solutions of pendulum equations (11.21) without control ($u = 0$) it follows, that if $A \neq 0$, $B \neq 0$ then equation (11.36) can be true if $H_0^1(p_1(t), q_1(t)) \equiv H_0^2(p_2(t), q_2(t))$ or pendulums are in equilibrium states[3].

Assumed that control goal (11.15) is not fulfilled and $|A| \neq |B|$. Then the equation (11.36) is true only when pendulums are in equilibrium states. The inequality $|A| \neq |B|$ is always true for the area $H_0^2(p_2, q_2) < E_2$, when

$$\alpha > E_2^2 max\{E_1, E_2 - E_1\}. \qquad (11.37)$$

---

[3]The linear combination of periodic functions can be zero only if periods are the same. The linear combination of $\sin q_1(t)$ and $\sin q_2(t)$ can not be zero if one pendulum makes circle oscillations and the other pendulum doesn't.

Consequently, if inequality (11.37) is true, then the solution of the closed-loop system achieves the control goal (11.15) or converges to an equilibrium state of uncontrolled system (11.21). There are only four equiliblium states

$$(q_1, p_1, q_2, p_2) = \{(0,0,0,0), (0,0,\pi,0), (\pi,0,0,0), (\pi,0,\pi,0)\}. \quad (11.38)$$

Analysis of equilibrium states. Matrix of linear approximation of the system (11.21), (11.27) near the state $(\pi,0,0,0)$ is

$$\begin{pmatrix} 0 & (ml^2)^{-1} & 0 & 0 \\ mgl & \Gamma A & 0 & -\Gamma B \\ 0 & 0 & 0 & (ml^2)^{-1} \\ 0 & -\Gamma A & -mgl & \Gamma B \end{pmatrix}. \quad (11.39)$$

Its characteristic polynomial is

$$l^2 x^4 - l^2\Gamma(B+A)x^3 + gl\Gamma(B-A)x - g^2 = 0. \quad (11.40)$$

Obviously, this equation has a positive root for any real A, B and an unstable manifold of the point $(\pi,0,0,0)$ is not trivial, according to the center manifold theorem, see [Khalil, 2002]. Therefore, initial conditions for the trajectories converging to this state should have zero projection onto the unstable manifold and the set of such initial conditions has zero Lebesgue measure.

Matrix of linear approximation of the system (11.21), (11.27) near the state $(0,0,\pi,0)$ is

$$\begin{pmatrix} 0 & (ml^2)^{-1} & 0 & 0 \\ -mgl & \Gamma A & 0 & -\Gamma B \\ 0 & 0 & 0 & (ml^2)^{-1} \\ 0 & -\Gamma A & mgl & \Gamma B \end{pmatrix}. \quad (11.41)$$

Its characteristic polynomial is

$$l^2 x^4 - l^2\Gamma(B+A)x^3 - gl\Gamma(B-A)x - g^2 = 0. \quad (11.42)$$

Obviously, this equation has a positive root for any real A, B and an unstable manifold of the point $(\pi,0,0,0)$ is not trivial, according to the center manifold theorem, see [Khalil, 2002]. Therefore, initial conditions for the trajectories converging to this state should have zero projection onto the unstable manifold and the set of such initial conditions has zero Lebesgue measure.

Matrix of linear approximation of the system (11.21), (11.27) near the state $(\pi, 0, \pi, 0)$ is

$$\begin{pmatrix} 0 & (ml^2)^{-1} & 0 & 0 \\ mgl & \Gamma A & 0 & \Gamma B \\ 0 & 0 & 0 & (ml^2)^{-1} \\ 0 & \Gamma A & mgl & \Gamma B \end{pmatrix}. \tag{11.43}$$

Its characteristic polynomial is

$$l^2 x^4 - l^2 \Gamma (B + A) x^3 - 2lgx^2 - gl\Gamma(B + A)x + g^2 = 0. \tag{11.44}$$

Obviously, this equation has a positive root for any real A, B and an unstable manifold of the point $(\pi, 0, 0, 0)$ is not trivial, according to the center manifold theorem, see [Khalil, 2002]. Therefore, initial conditions for the trajectories converging to this state should have zero projection onto the unstable manifold and the set of such initial conditions has zero Lebesgue measure.

Consequently, the following theorem is true.

Theorem 11.1. *Consider system (11.21) with feedback (11.27). Suppose that the assumptions (11.25, 11.26) are true. Then for any initial condition $(q_1(0), q_2(0), p_1(0), p_2(0))$ from set $\mathbb{X}_0 = \{(q_1, q_2, p_1, p_2) : V(q_1, q_2, p_1, p_2) < V(0, 0, 0, 0)\} \setminus \mathbb{D}$, where set $\mathbb{D}$ has zero Lebesgue measure, constraint condition (11.16) and control goal (11.15) fulfilled.*

Remark. Near the state $(0, 0, 0, 0)$ it is suggested to switch to another algorithm, which move the system from the area where $V(q_1, q_2, p_1, p_2) > V(0, 0, 0, 0)$ to the area where $V(q_1, q_2, p_1, p_2) < V(0, 0, 0, 0)$. For example a resonance control designed on one pendulum could be used. During simulations we never need such a switching.

Simulation. To demonstrate the ability of the controller to achieve the control goal and to fulfill the phase constraints we carried out computer simulation. The following value of system parameters and initial conditions were chosen: $m = 1$, $l = 1$, $g = 10$, $q_1 = 0$, $q_2 = 0.05$, $p_1 = 0$, $p_2 = 0$. Energy goal value for the first pendulum was taken $E_1 = 20$, energy constraint for the second one was taken $E_2 = 5$. Algorithm parameters were: $\Gamma = 0.015$, $\alpha = 10$. Time for simulating was 80 seconds. As predicted by control algorithm analysis the energy of the first pendulum converged to the goal value E_1, and energy of the second was constrainted by E_2. The simulating results are presented in Fig. 11.2, 11.3.

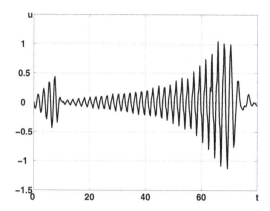

Fig. 11.2 Control function $u(q_1(t), p_1(t), q_2(t), p_2(t))$.

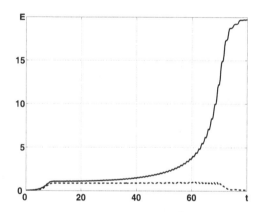

Fig. 11.3 Energy of pendulums. Solid line corresponds to the energy of the first pendulum $H_0^1(q_1(t), p_1(t))$, dash line — energy of the second one $H_0^2(q_2(t), p_2(t))$.

11.4 Conclusion

In the paper a method for control of mechanical systems under phase constraints, applicable to energy control of Hamiltonian systems is proposed. The constrained energy control problem for two pendulums by a single control action is studied both analytically and numerically. It is shown that for a proper choice of penalty parameter of the algorithm any energy level for the one pendulum under specified constraint on the energy of the other pendulum can be achieved. Simulation results confirm fast convergence rate of the algorithm.

Future research is aimed on application of the method to constrained energy control of Huijgens' pendula system [Pogromsky *et al.*, 2006].

Acknowledgment

The authors were supported by NWO-RFFI 047.011.2004.004, RFFI 08-01-00775 and the Council for grants of the RF President to support young Russian researchers and leading scientific schools (project NSh-2387.2008.1).

References

Fradkov, A. L. (1979), Speed-gradient scheme and its applications in adaptive control. *Autom. Remote Control*, 40:1333–1342.

Fradkov, A. L. (1990), *Adaptive Control in Complex Systems*. Nauka. Moscow. (in Russian).

Fradkov, A. L., Pogromsky, A. Yu. (1998), *Introduction to Control of Oscillations and Chaos*. World Scientific. Singapore.

Fradkov, A. L., Miroshnik, I. V., Nikiforov, V. O. (1999), *Nonlinear and adaptive control of complex systems*. Kluwer Academic Publishers. Dordrecht.

Fradkov, A. L. (2007), *Cybernetical physics*. Springer-Verlag. Berlin Heidelberg.

Fiacco, A. V., McCormick, G. P. (1990), *Nonlinear Programming: Sequential Unconstrained Minimation Techniques*. 2nd edn. SIAM.

Filippov, A. F. (1988). *Differential Equations with Discontinuous Righthand Sides*. Springer.

Nijmeijer, H, van der Schaft, A. J. (1990), *Nonlinear dynamical control systems*. Springer-Verlag. New York.

Tomchina, O. P., Tomchin, D. A., Fradkov, A. L. (2005), Speed-gradient control of passing through resonance in one-and two-dimensional motion. *16th IFAC World Congress Autom. Control.* June.

Anan'evskii, M. S. (2007) Selective control of the observables in the ensemble of quantum-mechanical molecular systems. *Autom. Remote Control*, 68:1333–1342.

Khalil, H. K. (2002), *Nonlinear systems*. 3rd edn, Prentice-Hall, Upper Saddle River.

Pogromsky, A. Yu, Belykh, V. N., Nijmeijer, H. (2006), A study of controlled synchronization of Huijgens' pendula. In K. Y. Pettersen, J. T. Gravdahl, H. Nijmeijer, editors, *Group coordination and cooperative control*, volume 336, pages 205–216. Springer-Verlag Berlin Heidelberg.

Two Van der Pol-Duffing Oscillators with Huygens Coupling

V.N. Belykh*, E.V. Pankratova*, A.Y. Pogromsky†

*Volga State Academy of Water Transport,
Nizhniy Novgorod,
Russia

†Department of Mechanical Engineering,
Eindhoven University of Technology,
5600 MB, Eindhoven, The Netherlands

Abstract

We consider a system of two Van der Pol-Duffing oscillators with Huygens (speeding up) coupling. This system serves as appropriate model for Huygens synchronization of two mechanical clocks hanging from a common support. We examine the main regimes of complete and phase synchronization, and study the dependence of their onset on the initial conditions. In particular, we reveal co-existence of two chaotic phase synchronized modes and study the structure of their complicated riddled basins.

12.1 Introduction

Synchronization phenomena have been the subject of discussion in various research areas since the 17th century, when the synchronization of two pendulum clocks attached to a common support beam was first discovered by Cristian Huygens [Huygens, 1673]. Later, the findings of Huygens were reproduced by means of similar setups in both experimental and theoretical works [Blekhman, 1998; Bennett *et al.*, 2002; Pantaleone, 2002].

In a recent paper [Oud *et al.*, 2006] synchronization experiments with a setup consisting of driven pendula were performed. Particular attention was paid to different synchronization regimes that can be observed in this situation: anti-phase and in-phase synchronization which are two typical synchronization regimes observed in the experimental setup. Historically, anti-phase synchrony was originally observed by Huygens while in-phase synchronization in a Huygens-type setup was observed and explained in [Blekhman, 1998] by means of a model of coupled Van der Pol equations.

Here, we examine a model of Huygens original system that consists of two oscillators driven by the Van der Pol-Duffing control input, providing the escape mechanism. Both oscillators are connected to a movable platform via a spring. Due to this coupling the oscillators influence each other and can synchronize. This paper investigates the onset of synchronous regimes in the two-oscillator setup and their dependence on the initial conditions and parameters of the coupled system.

12.2   Problem statement

The mathematical mechanism behind the remarkable Huygens' observation of synchronized mechanical clocks hanging from a common support was recently described in [Pogromsky *et al.*, 2003]. The model used is the following. A beam of mass M can move in the horizontal direction with viscous friction defined by damping coefficient d. One side of the beam is attached to the wall via a spring with elasticity k. The beam supports two identical pendula of length l and mass m. The torque applied to each pendulum is to sustain the clock active mode of pendula. It can also be viewed as a control input, introducing the escape mechanism to the system. The system equations can be written in the form of Euler-Lagrange equations:

$$
\begin{aligned}
ml^2\ddot{\phi}_1 + mgl\sin\phi_1 + f_1 &= -ml\ddot{y}\cos\phi_1, \\
ml^2\ddot{\phi}_2 + mgl\sin\phi_2 + f_2 &= -ml\ddot{y}\cos\phi_2, \\
(M+2m)\ddot{y} + d\dot{y} + ky &= -ml\sum_{i=1}^{2}(\ddot{\phi}_i\cos\phi_i - \dot{\phi}_i^2\sin\phi_i),
\end{aligned}
\tag{12.1}
$$

where $\phi_i \in S^1$ is the angular displacement of the i-th pendulum about its pivot point, y is the linear displacement of the beam, and f_1, f_2 are the control inputs defining the escape mechanism.

The Hamiltonian for each unperturbed pendulum ($f_i \equiv 0, y \equiv 0$) has the form:

$$H(\phi_i, \dot{\phi}_i) = \frac{ml^2 \dot{\phi}_i^2}{2} + mgl(1 - \cos \phi_i), \tag{12.2}$$

Here, in contrast to [Pogromsky *et al.*, 2003], we use the only angle dependent control input

$$f_i = \gamma \dot{\phi}_i [H(\phi, 0) - H_*], \tag{12.3}$$

and get a simple model of the pendulum

$$ml^2 \ddot{\phi} + \gamma \dot{\phi} [H(\phi, 0) - H_*] + mgl \sin \phi = 0. \tag{12.4}$$

Defined on the cylinder ($\phi, \dot{\phi}$), system (12.4) has an unstable equilibrium point ($\phi = \dot{\phi} = 0$), enveloped by a stable limit cycle which turns into a heteroclinic contour of the saddle point ($\phi = \pi, \dot{\phi} = 0$) when parameter $H_* > 0$ increases [Belyustina and Belykh, 1973]. Since the clock angular displacement is fairly small (there is no rotation), the parameter H_* must be small, and one can pass to the linear displacement of pendula $x_i = l\phi_i$ using shortened expansions in (12.1), (12.4).

$$
\begin{array}{ll}
\cos \phi = 1 - \frac{x^2}{2l^2} + \dots, & \sin \phi = \frac{x}{l} - \frac{x^3}{6l^3} + \dots, \\
H(\phi, 0) = mg\frac{x^2}{2l} + \dots, & \ddot{y} \cos \phi = \ddot{y} + \dots, \\
\ddot{\phi} \cos \phi = \ddot{\phi} + \dots, & \dot{\phi}^2 \sin \phi = 0.
\end{array}
\tag{12.5}
$$

Obviously, this expansion is valid for a large enough values of l. Hence, the system (12.1) attains the form:

$$
\begin{array}{l}
\ddot{x}_1 + \lambda(x_1^2 - 1)\dot{x}_1 + \omega^2 x_1 - \alpha x_1^3 = -\ddot{y}, \\
\ddot{x}_2 + \lambda(x_2^2 - 1)\dot{x}_2 + \omega^2 x_2 - \alpha x_2^3 = -\ddot{y}, \\
(M + 2m)\ddot{y} + d\dot{y} + ky = -m(\ddot{x}_1 + \ddot{x}_2),
\end{array}
\tag{12.6}
$$

where $\lambda = \frac{\gamma g}{2l^3}$, $\omega = \sqrt{g/l}$, $\alpha = \frac{g}{6l^3}$.

The system (12.6) can be rewritten in the following form

$$
\begin{array}{l}
\ddot{x}_1 + \omega^2 x_1 + F(x_1, \dot{x}_1) = -\mu\ddot{y}, \\
\ddot{x}_2 + \omega^2 x_2 + F(x_2, \dot{x}_2) = -\mu\ddot{y}, \\
\ddot{y} + h\dot{y} + \Omega^2 y = m[\omega^2(x_1 + x_2) + \sum\limits_{i=1}^{2} F(x_i, \dot{x}_i)], \\
\dot{x}_1 = u_1, \dot{x}_2 = u_2, \dot{y} = z.
\end{array}
\tag{12.7}
$$

Here, $F(x_i, \dot{x}_i) = \lambda(x_i^2 - 1)\dot{x}_i - \alpha x_i^3, i = 1, 2$, and new notations for the coupling parameter $\mu = 1/M$, frequency of the platform $\Omega = \sqrt{k/M}$ and

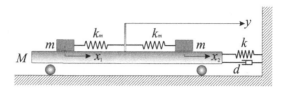

Fig. 12.1   Schematic drawing of the system, which consists of two oscillators with masses m coupled via the platform with mass M.

damping factor $h = d/M$ were introduced. Note that new y stands for μy. Thus, the Huygens problem is now transformed to the synchronization problem of two Van der Pol-Duffing oscillators with the special coupling which we call "Huygens coupling". Note that for $\alpha = 0$ and $\lambda \ll 1$ such an approach was used in [Blekhman, 1988].

For the linear displacement x_i the equations (12.7) model the Nijmeijer setup described in [Oud et al., 2006] and depicted in Fig.12.1. The platform of the mass M is attached to the wall via a spring with linear stiffness k and a damper with viscous friction coefficient d. Due to its possible horizontal displacement the platform couples the dynamics of two oscillators with masses m. Both oscillators are connected to the platform via a spring with linear stiffness k_m and driven by force $F(x_i, \dot{x}_i) = \lambda(x_i^2 - 1)\dot{x}_i - \alpha x_i^3, i = 1, 2$. In the setup shown in Fig.12.1 the frequency of each m-mass oscillator is $\omega = \sqrt{k_m/m}$. In Eq.(12.7) $x_{1,2}$ and y are the vibrations of oscillators and platform, respectively; $u_{1,2}$ and z are the velocities of motion of the oscillators and platform, respectively.

The conservative system (12.6), with $\lambda = d = 0$ has the integral

$$V = \sum_{i=1}^{2} m\left(\frac{\omega^2 x_i^2}{2} - \frac{\alpha x_i^4}{4} + \frac{\dot{x}_i^2}{2} + \dot{x}_i \dot{y}\right) + \\ (M + 2m)\frac{\dot{y}^2}{2} + k\frac{y^2}{2} = const \tag{12.8}$$

which serves as the Lyapunov-like function for the system with the damped platform, $\lambda = 0, d > 0$:

$$\dot{V} = -d\dot{y}^2 \leq 0.$$

The latter implies that conservative oscillators together with the platform for $d > 0$ return to the stable equilibrium point at the origin, provided that initial perturbations are bounded.

Our goal is to study synchronization phenomena when the escape mechanism is switched on, i.e. $\lambda > 0$. From the equations of the model it follows that the system (12.6), as well as the system (12.7), has two invariant manifolds: the 4-d in-phase manifold $M_s := \{(x_1, u_1) = (x_2, u_2)\}$ and the 2-d anti-phase manifold $M_a := \{(x_1, u_1) = (-x_2, -u_2), y = z = 0\}$. The system (12.7) possess two symmetries defining the existence of the two manifolds: the invariance under the maps

$$\begin{aligned} (x_1, x_2) &\rightarrow (x_2, x_1) \\ (u_1, u_2) &\rightarrow (u_2, u_1) \end{aligned} \tag{12.9}$$

and

$$\begin{aligned} (x_1, x_2, y) &\rightarrow (-x_1, -x_2, -y) \\ (u_1, u_2, z) &\rightarrow (-u_1, -u_2, -z) \end{aligned} \tag{12.10}$$

The map (12.9) defines the mirror symmetry with respect to M_s, while the transformation (12.10) gives the central symmetry with respect to M_a.

From existence of manifolds, and from symmetries (12.9), (12.10) it follows that the system (12.7) may have a synchronous attractor A_s lying in M_s, an anti-phase attractor A_a contained in M_a. Besides A_s and A_a, there may exist either a symmetrical attractor A_{sym} lying outside of $(M_s \cup M_a)$ or two asymmetrical attractors A_{asym}^+, A_{asym}^- being symmetrical to each other via (12.9), (12.10). Obviously, the attractors may coexist in the phase space of the system (12.7). Our main purpose is to study these attractors and the structure of their basins.

12.3 Synchronization of oscillators driven by Van der Pol control input

We start with the case where both oscillators are driven by the force $F(x_i, \dot{x}_i) = \lambda(x_i^2 - 1)\dot{x}_i, i = 1, 2$. Thus, for $\mu = 0$, the behavior of each oscillator is defined by a solution of the differential equation, known as the Van der Pol equation. The parameter λ defines the form of the limit cycle: for small λ the motion of oscillator is quasi-harmonic, for large λ the relaxation oscillations are observed. We consider the case, where the frequencies of the oscillators and platform are equal. The similar situation was considered in [Oud et al., 2006], where the experiments with the frequency of the platform which is close to or above the frequency of the metronomes have been performed. We assume that $\omega = \Omega = 1$. Experimentally, to prevent the situation when oscillations of the platform become too large and the metronomes will hit the frame of the setup, the authors of [Oud et al., 2006] used (increased) magnetic damping. In our theoretical study we assume that $h = 0.5$.

For $\mu \neq 0$ the platform couples both oscillators. Since the coupling parameter μ is inversely proportional to the mass of the platform M, a decrease in M leads to an increase in μ. In his original experiment, Huygens used a massive beam to couple the clocks and found that phase locking in anti-phase regime was robust [Huygens, 1673]. However, it was shown later, that the synchronization of two metronomes resting on a light wooden board [Pantaleone, 2002] generally appeared in the form of an in-phase motion. Both the situations can easily be modeled and confirmed in our setting as limit cases for small and large μ, respectively. In the present

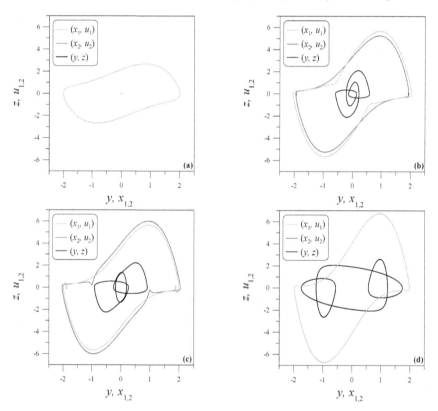

Fig. 12.2   Projections of the phase portrait onto the planes (x_1, u_1), (x_2, u_2), and (y, z) for various values of parameter λ. For all the curves initial conditions are the same $(x_1^0, x_2^0) = (0.4, -0.1)$. (a) $\lambda = 1$; (b) $\lambda = 2.64$; (c) $\lambda = 2.8$; (d) $\lambda = 3$. Parameters are: $\alpha = 0, \Omega = \omega = 1, \mu = 0.5, h = 0.5, m = 0.47$.

work, we examine the intermediate case and consider the two following values for the coupling parameter: $\mu = 0.5$ and $\mu = 1$. Let us first focus on the case where $\mu = 0.5$. For various λ the limit sets for the trajectory starting from the phase point $(x_1^0, x_2^0) = (0.4, -0.1)$ are shown in Fig.12.2 in projections onto the (displacement, velocity)-planes. The initial values for other variables are assumed to be zero.

Obviously, that for identical oscillators in anti-phase regime the total force would be the zero and due to the damper, oscillations of the platform die out, Fig12.2(a). When the dynamics of oscillators is on attractors A_{asym}^+ or A_{asym}^-, Fig12.2(b), (c) as well as for the completely synchronous movement of both oscillators observed in in-phase regime, Fig12.2(d), oscillations of the platform appear.

The basins of attractors corresponding to these oscillatory regimes are shown in Fig.12.3 in projection onto the (x_1, x_2)-plane. In order to obtain these diagrams, we have changed initial displacements of both oscillators within the range $X := \{(x_1^0, x_2^0) | x_1^0 \in [-3, 3], x_2^0 \in [-3, 3]\}$. Initial values for other variables were assumed to be equal to zero.

For small λ both oscillators only oscillate in either anti-phase or in-phase modes, Fig.12.3(a), (b). In the figures the anti-phase synchronization $(A_a \subset M_a := \{(x_1, u_1) = (-x_2, -u_2), y = z = 0\})$ is established for initial oscillators' displacements falling into the light gray ranges. Basins of attraction corresponding to the in-phase synchronization $(A_s \subset M_s := \{(x_1, u_1) = (x_2, u_2)\})$ are shown in black. For $\lambda \gtrsim 2.63$ for some initial conditions the dynamics of oscillators can also obey various regular symmetrical to each other attractors A_{asym}^+ or A_{asym}^-. The basins of these attracting sets are shown as gray and dark gray domains in Fig.12.3(c), (d). For large enough λ, attractor A_a belonging to the manifold M_a and corresponding to oscillations in anti-phase regime, loses its stability. In this case three stable periodic orbits co-exist in phase space of the system (12.7): A_{asym}^+, A_{asym}^- and A_s, belonging to the diagonal M_s. For $\lambda = 3$ the basins of these attractors in projections onto the (x_1, x_2) plane are shown in Fig.12.3(d). Note that the structures shown in Fig.12.3 were obtained for zero values of initial oscillators' velocities. Otherwise, if $u_1^0 \neq u_2^0$, the symmetry observed in Fig.12.3 is broken.

When $\mu = 1$ for small and large values of λ the structures of attractors similar to Fig.12.3(a), (d) are observed. However, the transition from the double-state $\{A_s$ and $A_a\}$ to the triple-state $\{A_s, A_{asym}^+$ and $A_{asym}^-\}$ regime occurs differently. To illustrate this we consider a one-parameter bifurcation diagram for increasing λ, obtained for the fixed initial displacements of oscillators: $(x_1^0, x_2^0) = (-0.7, -2.0)$, Fig.12.4. In this figure, x_2 is the value of displacement of the second oscillator being the point of intersection of the

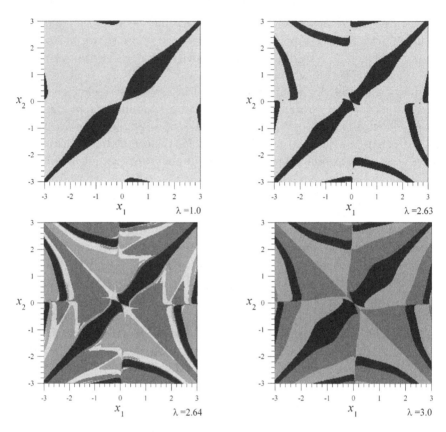

Fig. 12.3   Basins of attraction in projections onto the (x_1, x_2) plane for various values of parameter λ. The domains for anti-phase synchronization $((x_1, u_1) = (-x_2, -u_2))$ are shown in light gray, whereas the black domains correspond to in-phase synchronization $((x_1, u_1) = (x_2, u_2))$. The basins of attractors A^+_{asym} and A^-_{asym} are shown in gray and dark gray, respectively. The parameters of the system are: $\alpha = 0, \Omega = \omega = 1, \mu = 0.5, h = 0.5, m = 0.47$.

trajectory with the hyperplane $\mathcal{P} : \{x_1 = 0.1\}$. As seen from Fig.12.4, the transition from double-state to triple-state behavior for $\mu = 1$ via chaotic regime occurs. Experimentally, the existence of such type of oscillations was also observed in [Oud *et al.*, 2006]. In Fig.12.5(a) the projection of a chaotic attractor onto the (x_1, x_2) plane for $\lambda = 3$ is presented. When $\lambda = 3.2$, as for $\mu = 0.5$, two regular, symmetrical to each other attractors A^+_{asym} or A^-_{asym} are observed, Fig.12.5(b).

In order to show existence of phase synchronization for oscillators, we calculate the phases of their oscillations. In our case, since the center of

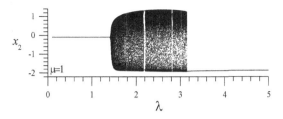

Fig. 12.4 Rearrangement of attracting sets on the one-parametric bifurcation diagram for $\mu = 1$. The initial displacements of oscillators are $(x_1^0, x_2^0) = (-0.7, -2.0)$.

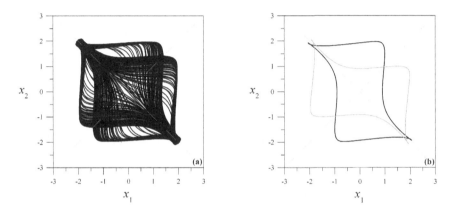

Fig. 12.5 Projections of phase portrait onto the (x_1, x_2) plane for $\mu = 1$ and (a) $\lambda = 3$, (b) $\lambda = 3.2$.

rotation can clearly be distinguished at $(x_i, u_i) = (0, 0), i = 1, 2$, Fig.12.2, the following definition of the phase can be used:

$$\theta_{1,2} = \arctan(u_{1,2}/x_{1,2}). \qquad (12.11)$$

For various values of parameters we check whether the locking condition

$$\Delta\theta = |\theta_1 - \theta_2| < const < 2\pi \qquad (12.12)$$

is satisfied. In Fig.12.6 the phase difference $\Delta\theta$ for regular ($\lambda = 3.2$) and chaotic ($\lambda = 3$) motions of oscillators is given. For both the cases $\Delta\theta$ is a bounded function. Accordingly, phase synchronization is observed for both regular and chaotic motions in the considered setup.

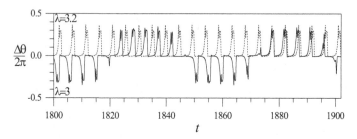

Fig. 12.6   Phase difference $\Delta\theta$ for $\lambda = 3$ - solid line, and $\lambda = 3.2$ - dotted line, $\mu = 1$.
Here, $\Delta\theta < \pi$.

12.4   Synchronization of oscillators driven by Van der Pol-Duffing control input

In this section we assume that both oscillators are driven by the force $F(x_i, \dot{x}_i) = \lambda(x_i^2 - 1)\dot{x}_i - \alpha x_i^3$, $i = 1, 2$, and focus on changes in the behavior of oscillators occurring with the increase of α. Depending on the parameters, the structure of the phase space for individual oscillator ($\mu = 0$) can be different [Belyustina and Belykh, 1973]. For small values of α the unstable trivial equilibrium point $x^* = 0$ is enveloped by a stable limit cycle. The influence of two saddle points $x^* = \pm\omega/\sqrt{\alpha}$ in this case can be neglected. With the increase of α both saddle points move to each other resulting in formation of a heteroclinic contour of saddle points for some value $\alpha = \alpha_h$. While α reaches α_h the limit cycle "grows into" this heteroclinic contour. For $\alpha > \alpha_h$ the limit cycle disappears.

For the system (12.7) there are nine equilibrium points. All of these states are of saddle type. The structure of basins for attractors in system (12.7) under Van der Pol-Duffing control input for various values of parameter α and fixed other parameters ($\lambda = 2.5, \Omega = 0.3, \omega = 1, \mu = 1, h = 0.01, m = 0.47$) is shown in Fig.12.7. Note that similar structures for parameters considered in the previous section, with increasing α can also be observed. For small α three various attracting sets co-exist in the system: $A_s \subset M_s$, and two symmetrical to each other attractors A_{asym}^+ and A_{asym}^-, Fig.12.7(a). The increase of α leads to disappearance (at $\alpha = \alpha^* \approx 0.15$) of attractors A_{asym}^+ and A_{asym}^-, Fig.12.7(b). For $\alpha = \alpha^*$ the manifolds of nontrivial equilibrium points divide the phase space of the system into two parts: the part of completely synchronous motion $S : \{(x_1, x_2)||x_1| < \omega/\sqrt{\alpha^*}, |x_2| < \omega/\sqrt{\alpha^*}\}$ (black range in Fig.12.7) and the part of unstable motion $U : \{(x_1, x_2)||x_1| > \omega/\sqrt{\alpha^*}, |x_2| > \omega/\sqrt{\alpha^*}\}$.

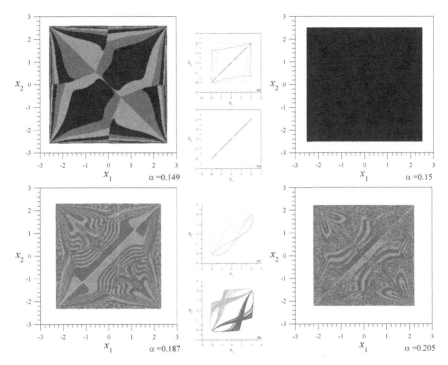

Fig. 12.7   Basins of attraction in projections onto the (x_1, x_2) plane for various values of parameter α. The in-phase synchronization $((x_1, u_1) = (x_2, u_2))$ is established for initial displacements of oscillators falling into the black ranges. The ranges shown in gray and dark gray correspond to the case of phase synchronization when the dynamics of each oscillator is defined by various, symmetrical to each other attractors A_{asym}^{+} and A_{asym}^{-}. The parameters of the system (12.7) are: $\lambda = 2.5, \Omega = 0.3, \omega = 1, \mu = 1, h = 0.01, m = 0.47$.

This regime is preserved for $\alpha \in (0.15; 0.18)$. For $\alpha \approx 0.187$ the attractor A_s loses its asymptotic transverse stability via a *riddling bifurcation* [Lai *et al.*, 1996]. After the riddling bifurcation, the system has two new stable periodic solutions $\widetilde{A}_{asym}^{+}$ and $\widetilde{A}_{asym}^{-}$, Fig.12.7(c). For $\alpha = 0.205$ the complex (riddled) structure of basin for chaotic attractors is shown in Fig.12.7(d). The phase portrait of the system in projections onto the (x_1, u_1) and (y, z) planes for $\alpha = 0.205$ is shown in Fig.12.8(a), (b). The following increase of α leads to the appearance of a more complicated chaotic attractor, Fig.12.8(c), (d).

To illustrate rearrangement of attracting sets in the system (12.7) we consider a one-parameter bifurcation diagram for increasing α, Fig.12.9. In

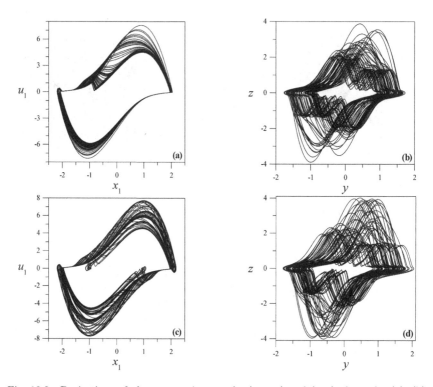

Fig. 12.8   Projections of phase portrait onto the (x_1, u_1) and (y, z) planes for (a), (b) $\alpha = 0.205$ and (c), (d) $\alpha = 0.21$.

this figure, as before, x_2 is the value of displacement for the second oscillator being the point of intersection of the trajectory with the hyperplane $\mathcal{P}$: $\{x_1 = 0\}$. The initial conditions were taken close to the diagonal M_s, namely $(x_1^0, x_2^0) = (0.0, -0.1)$. Therefore, for $0 < \alpha < 0.187$ the behavior of both oscillators are completely synchronous: x_1 and x_2 intersect $\mathcal{P}$ at the same time. For $0.187 < \alpha < 0.202$, depending on the initial conditions we observe two various values of displacement for the second oscillator (it is omitted in Fig.12.9 for brevity). In this interval of α two various but symmetrical to each other periodic motions $\widetilde{A}_{asym}^{+}$ and $\widetilde{A}_{asym}^{-}$ co-exist. The chaotic oscillations arise at $\alpha \approx 0.203$ and preserve in a wide range of α. However, despite this complexity of motion, both oscillators exhibit the phase synchronous behavior.

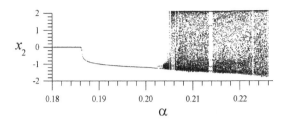

Fig. 12.9 Transition to chaos in the one-parametric bifurcation diagram with the increase of α. The initial displacements of oscillators are $(x_1^0, x_2^0) = (0.0, -0.1)$.

12.5 Conclusions

In the present work we have considered the system of two oscillators connected to a movable platform via a spring. The platform couples the dynamics of both oscillators causing their synchronous movement. In order to analyze possible regimes of their synchronous motion the numerical simulation has been performed for two types of control input: Van der Pol and Van der Pol-Duffing. It was shown that depending on the initial conditions and parameters of the system, three synchronous regimes are possible: complete $((x_1, u_1) = (x_2, u_2))$, anti-phase $((x_1, u_1) = (-x_2, -u_2))$ and phase synchronization. For the latter regime the amplitudes of two oscillators remain quite different, while the locking between the phases has been observed, i.e. $\Delta\theta = |\theta_1 - \theta_2| < const < 2\pi$. The existence of this regime for both regular and irregular motions in the setup has been revealed. Phase synchronization implies that despite both the complexity of motion and difference in oscillator displacements the clocks show the same time.

Acknowledgments

This work was supported in part by the Russian Foundation for Basic Research (grants No. 09-01-00498 and No. 07-02-01404) and the Dutch-Russian program Dynamics and Control of Hybrid Mechanical Systems (DyCoHyMS) (NWO-RFBR grant 047.017.018). E.V.P. also acknowledges the support of the Dynasty Foundation.

References

Belyustina, L.N. and Belykh, V.N. (1973) Qualitative analysis of dynamical system on cylinder. *Differents. Uravneniya*, **9**(3), pp. 403–415.

Bennett, M., Schatz, M., Rockwood, H. and Wiesenfeld, K. (2002) Huygenss clocks. *Proc. R. Soc. Lond. A*, **458**(2019), pp. 563-579.

Blekhman, I.I. (1998). *Synchronization in science and technology*. ASME. New York.

Huygens, C. (1673) *Horoloquim Oscilatorium*. Apud F. Muguet, Parisiis, France; English translation: *The pendulum clock*, 1986, Iowa State University Press, Ames.

Lai, Y.-C., Grebogi, C., Yorke, J.A. and Venkataramani, S.C. (1996) Riddling bifurcation in chaotic dynamical systems. *Phys. Rev. Lett.*, **77**, pp. 55-58.

Oud, W.T., Nijmeijer, H. and Pogromsky, A.Yu. (2006) A study of Huijgens synchronization. Experimental results, in *Group Coordinationi and Control*, K.Y. Pettersen, J.T. Gravdahl, H. Nijmeijer (eds), Springer, 2006.

Pantaleone, J. (2002) Synchronization of metronomes. *American Journal of Physics*, **70**(10), pp. 992-1000.

Pogromsky, A.Yu., Belykh, V.N. and Nijmeijer, H. (2003). Controlled synchronization of pendula. In *Proc. of the 42nd IEEE Conference on Decision and Control*. Maui, Hawaii USA, December . pp. 4381–4386.

Pogromsky, A.Yu., Belykh, V.N. and Nijmeijer, H. (2006). A study of controlled synchronization of Huijgens pendula. In *Lecture Notes in Control and Information Sciences. Group Coordination and Cooperative Control.*, **336**, pp. 205–216. Springer Berlin. Heidelberg.

Van der Pol, B. (1922) Theory of the amplitude of free and forced triod vibration. *Radio Rev.*, **1**, pp. 701-710.

Synchronization of Diffusively Coupled Electronic Hindmarsh-Rose Oscillators

E. Steur, L. Kodde, H. Nijmeijer

Department of Mechanical Engineering
Eindhoven University of Technology
P.O. Box 513, 5600 MB Eindhoven, The Netherlands

Abstract

In this chapter we present results on synchronization of diffusively coupled Hindmarsh-Rose (HR) electronic oscillators. These electronic oscillators are analog electrical circuits which integrate the differential equations of the Hindmarsh-Rose model. An experimental setup consisting of four chaotic Hindmarsh-Rose oscillators, is used to evaluate the existence and stability of partially or fully synchronized states.

13.1 Introduction

Synchronous behavior of systems is witnessed in a vast number of research area's. Beautiful examples are, for instance, the simultaneous flashing of male fireflies on banks along rivers in Malaysia, Thailand and New Guinea [Strogatz and Stewart, 1993], and the synchronous release of action potentials in parts of the mammalian brain [Gray, 1994]. A large number of examples of synchronization in nature can be found in [Pikovsky *et al.*, 2003]. Synchronization can also be found in robotics, usually referred to as coordination [Rodriguez-Angeles and Nijmeijer, 2001], and it is potentially of interest in the field of (secure) communication, see for instance [Pecora and Carroll, 1990; Huijberts *et al.*, 1998].

Most of the studies on synchronization in networks of coupled systems deal with analysis supported by simulations. Significantly less attention is given to validate synchronization in an experiment setup. In this chapter we present synchronization of coupled Hindmarsh-Rose electronic oscillators. The Hindmarsh-Rose model [Hindmarsh and Rose, 1984] is a well-known model in the field of neuroscience that provides a description of the action potential generation in neuronal cells. This model consists of three coupled nonlinear differential equations and is capable, as function of specific parameters, of producing both simple and complex oscillatory motion. Here, an experimental setup consisting of four circuits, operating in a chaotic regime, is used to investigate the existence and stability of synchronized states.

The Hindmarsh-Rose oscillators in the experimental setup are diffusively coupled; that is the systems are mutually coupled using linear functions of the outputs of the systems. Using a semipassivity based approach [Pogromsky *et al.*, 2002], we derive conditions that guarantee the existence of synchronized regimes. These regimes might correspond to the fully synchronized state, i.e. all systems perform an identical motion, as well as to partial synchronization where only some systems do synchronize.

This chapter is organized as follows. In section 13.2 the mathematical notations are being introduced and we present the notions of *semipassivity* and *convergent systems*. In section 13.3 we present a theoretical passivity-based framework, introduced in [Pogromsky, 1998], that provides conditions under which the coupled oscillators synchronize. In addition, we show that the Hindmarsh-Rose systems satisfy the assumptions of this framework. Next, in section 13.4 the experimental setup is discussed and in section 13.5 we show that the coupled electronic Hindmarsh-Rose systems synchronize. Finally, in section 13.6 conclusions are drawn.

13.2   Preliminaries

Throughout this paper we use the following notations. The Euclidian norm in $\mathbb{R}^n$ is denoted by $\| \cdot \|$, $\|x\|^2 = x^\top x$ where the symbol $^\top$ stand for transposition. The symbol I_n defines the $n \times n$ identity matrix and the notation $\mathrm{col}\,(x_1, \ldots, x_n)$ stands for the column vector containing the elements $x_1, \ldots, x_n$. A function $V : \mathbb{R}^n \to \mathbb{R}_+$ is called positive definite if $V(x) > 0$ for all $x \in \mathbb{R}^n \setminus \{0\}$ and $V(0) = 0$. It is radially unbounded if $V(x) \to \infty$ if

$\|x\| \to \infty$. If the quadratic form $x^\top P x$ with a symmetric matrix $P = P^\top$ is positive definite, then the matrix P is positive definite, denoted as $P > 0$. The notation $A \otimes B$ stands for the Kronecker product of the matrices A and B.

We define (practical) synchronization and (practical) partial synchronization as follows:

Definition 13.1 (synchronization). *Consider k interconnected dynamical systems with state variables $x_i \in \mathbb{R}^n$, $i = 1, 2, \ldots, k$. The coupled systems are called*

- synchronized *if* $\lim_{t \to \infty} \|x_i(t) - x_j(t)\| = 0$ *for all* $i, j = 1, 2, \ldots, k$;
- practically synchronized *if* $\limsup_{t \to \infty} \|x_i(t) - x_j(t)\| = \delta$ *for all* $i, j = 1, 2, \ldots, k$ *and some fixed, sufficiently small* $\delta > 0$;
- partially synchronized *if* $\lim_{t \to \infty} \|x_i(t) - x_j(t)\| = 0$ *for some* $i, j = 1, 2, \ldots, k$;
- practically partially synchronized *if* $\limsup_{t \to \infty} \|x_i(t) - x_j(t)\| = \delta$ *for some* $i, j = 1, 2, \ldots, k$ *and some fixed, sufficiently small* $\delta > 0$.

Let us, in addition, present the notions of *semipassivity* and *convergent systems*.

Definition 13.2 (semipassivity). *[Pogromsky and Nijmeijer, 2001] Consider the following system:*

$$\begin{aligned} \dot{x} &= f(x) + Bu \\ y &= Cx \end{aligned} \tag{13.1}$$

where state $x \in \mathbb{R}^n$, input $u \in \mathbb{R}^m$, output $y \in \mathbb{R}^m$, vector field $f : \mathbb{R}^n \to \mathbb{R}^n$ and matrices B and C of appropriate dimensions. Let $V : \mathbb{R}^n \to \mathbb{R}_+$, $V(0) = 0$ be a differentiable nonnegative (storage) function, then the system (13.1) is called semipassive if the following inequality is satisfied:

$$\dot{V} \le y^\top u - H(x) \tag{13.2}$$

where $H : \mathbb{R}^n \to \mathbb{R}$ is nonnegative outside some ball

$$\exists \rho > 0, \quad \forall \|x\| \ge \rho \Rightarrow H(x) \ge \varrho(\|x\|)$$

for some continuous nonnegative function $\varrho(\cdot)$ defined for $\|x\| \ge \rho$. If the inequality (13.2) is strict, the system (13.1) is called strictly semipassive.

The most useful property of semipassive systems is that being linearly interconnected, the solutions of all systems in the network exist for all $t \ge 0$ and are *ultimately* bounded [Pogromsky, 1998].

Consider the following system:

$$\dot{z} = q(z, w) \tag{13.3}$$

where $z(t) \in \mathbb{R}^s$, $w(t) \in \mathcal{D}$, $\mathcal{D}$ is some compact subset of $\mathbb{R}^p$, continuous function $w : \mathbb{R}_+ \to \mathcal{D}$ and the vector field $q : \mathbb{R}^s \times \mathcal{D} \to \mathbb{R}^s$.

Definition 13.3 (convergent systems). *[Demidovich, 1967; Pavlov et al., 2006] The system (13.3) is said to be convergent if for any $w(\cdot)$:*

(1) all solutions $z(t)$ are well-defined for all $t \in [t_0, +\infty)$ and all initial conditions $t_0 \in \mathbb{R}$, $z(t_0) \in \mathbb{R}^s$;

(2) there exists a unique globally asymptotically stable solution $z_w(t)$ defined and bounded for all $t \in (-\infty, +\infty)$, i.e. for any solution $z(t)$ it follows that

$$\lim_{t \to \infty} \|z(t) - z_w(t)\| = 0.$$

According to Demidovich [Demidovich, 1967], there exists a simple sufficient condition to determine if the system (13.3) is convergent.

Lemma 13.1. *[Demidovich, 1967; Pavlov et al., 2006] If there exists a matrix $P = P^\top > 0$ such that the eigenvalues $\lambda_i(Q)$ of the symmetric matrix*

$$Q(z, w) = \frac{1}{2} \left(P \frac{\partial q}{\partial z}(z, w) + \frac{\partial q}{\partial z}^\top (z, w) P \right) \tag{13.4}$$

are negative and separated away from the imaginary axis for all $z \in \mathbb{R}^s$, $w \in \mathcal{D}$, then the system (13.3) is convergent.

13.3 Synchronization of diffusively coupled Hindmarsh-Rose oscillators

In this section conditions are posed that guarantee (partial) synchronization in a network of coupled Hindmarsh-Rose oscillators. First the semipassivity based framework as described in [Pogromsky et al., 2002] is presented, and next we show that the Hindmarsh-Rose oscillators satisfy the assumptions of this framework.

Consider the k systems of the following form

$$\begin{aligned} \dot{x}_i &= f(x_i) + B u_i \\ y_i &= C x_i \end{aligned} \tag{13.5}$$

where $i = 1, 2, \ldots, k$ denotes the number of each system in the network, $x_i \in \mathbb{R}^n$ the state, $u_i \in \mathbb{R}^m$ the input and $y_i \in \mathbb{R}^m$ the output of the ith system, smooth vector field $f : \mathbb{R}^n \to \mathbb{R}^n$ and matrices B and C of appropriate dimensions. Let, in addition, the matrix CB be positive definite and nonsingular.

The k dynamical systems (13.5) are coupled via *diffusive* coupling, i.e. mutual interconnection through linear output coupling of the form

$$u_i = -\gamma_{i1}(y_i - y_1) - \gamma_{i2}(y_i - y_2) - \ldots - \gamma_{ik}(y_k - y_1) \tag{13.6}$$

where $\gamma_{ij} = \gamma_{ji} \geq 0$ denotes the strength of the interconnection between the systems i and j.

Defining the $k \times k$ *coupling matrix* as

$$\Gamma = \begin{bmatrix} \sum_{i=2}^{k} \gamma_{1i} & -\gamma_{12} & \cdots & -\gamma_{1k} \\ -\gamma_{21} & \sum_{i=1, i \neq 2}^{k} \gamma_{2i} & \cdots & -\gamma_{2k} \\ \vdots & \vdots & \ddots & \vdots \\ -\gamma_{k1} & -\gamma_{k2} & \cdots & \sum_{i=1}^{k-1} \gamma_{ki} \end{bmatrix}$$

the diffuse coupling functions (13.6) can be written as

$$\underline{u} = -\Gamma \underline{y} \tag{13.7}$$

where $\underline{u} = \mathrm{col}(u_1, \ldots, u_k)$, $\underline{y} = \mathrm{col}(y_1, \ldots, y_k)$. Since $\Gamma = \Gamma^\top$ all its eigenvalues are real. Moreover, applying Gerschgorin's theorem about the localization of the eigenvalues, it is easy to verify that Γ is positive semidefinite.

A network might possess certain symmetries. In particular, the network may contain repeating patterns. Hence, a permutation of some elements in the network, with respect to the interconnections, will leave the network unchanged. The mathematical representation of the permutation of the elements is a permutation matrix $\Pi \in \mathbb{R}^{k \times k}$. The matrix Π defines a symmetry for the network if Γ and Π commute, i.e. $\Pi\Gamma = \Gamma\Pi$. Moreover, given a permutation matrix Π that commutes with Γ, the set $\ker(I_{kn} - \Pi \otimes I_n)$ defines a *linear invariant manifold* for the closed loop systems (13.5) and (13.7). To be precise, the set $\ker(I_{kn} - \Pi \otimes I_n)$ describes a set of linear equations of the form $x_i - x_j = 0$ for some i and j. Hence, we want to guarantee asymptotic stability of such a set. Therefore, introduce a linear change of coordinates $x_i \mapsto (z_j, y_j)$. Under the assumption that CB is nonsingular, the systems (13.5) can be written after the coordinate transformation in the normal form:

$$\begin{cases} \dot{z}_i = q(z_i, y_i) \\ \dot{y}_i = a(z_i, y_i) + CBu_i \end{cases} \tag{13.8}$$

where $z_i \in \mathbb{R}^{n-m}$ and smooth vector fields $q : \mathbb{R}^{n-m} \times \mathbb{R}^m \rightarrow \mathbb{R}^{n-m}$, $a : \mathbb{R}^{n-m} \times \mathbb{R}^m \rightarrow \mathbb{R}^m$. A sufficient condition for asymptotic stability of the set $\ker(I_{kn} - \Pi \otimes I_n)$ is given in the following theorem:

Theorem 13.1. *[Pogromsky et al., 2002] Let λ' be the smallest nonzero eigenvalue of Γ under the restriction that the eigenvectors are taken from the set range $(I_k - \Pi)$. Assume that:*

A1. the free system (13.8) is strictly semipassive with respect to input u_i and output y_i with a radially unbounded storage function;

A2. there exists a matrix $P = P^\top > 0$ such that the conditions of Lemma 13.1 are satisfied for q as defined in (13.8).

Then for all positive semidefinite matrices Γ all solutions of the diffusive *network (13.8) and (13.7) are ultimately bounded and there exists a positive number $\bar{\lambda}$ such that if $\lambda' \geq \bar{\lambda}$ the set $\ker(I_{kn} - \Pi \otimes I_n)$ contains a globally asymptotically stable subset.*

A network of Hindmarsh-Rose oscillators is given by the following set of equations:

$$\begin{cases} \dot{y}_i = -a_1 y_i^3 + a_2 y_i + a_3 + a_4 z_{1,i} - a_5 z_{2,i} + v + u_i \\ \dot{z}_{1,i} = -b_1 - b_2 y_i^2 - b_3 y_i - b_4 z_{1,i} \\ \dot{z}_{2,i} = c_1 (c_2 (y_i + c_3) - z_{2,i}) \end{cases} \tag{13.9}$$

where $\dot{} := \frac{d}{d\tau}$, $\tau = 1000t$, $i = 1, 2, \ldots, k$ denotes the number of each oscillator in the network, y_i represents the membrane potential, which can be regarded as the natural output of a neuron, $z_{1,i}$ is an internal recovery variable and $z_{2,i}$ is a slow internal recovery variable and input u_i. The constant v represents an external applied stimulus. Parameters $a_1, a_2, a_3, a_4, a_5, b_1, b_2, b_3, b_4, c_1, c_2, c_3$ are all positive constants.

Lemma 13.2. *Each free Hindmarsh-Rose system is strictly semipassive with respect to the input u_i and the output y_i with a radially unbounded storage function.*

Proof. Following [Oud and Tyukin, 2004], consider the following storage function $V : \mathbb{R}^3 \rightarrow \mathbb{R}_+$:

$$V(y_i, z_{1,i}, z_{2,i}) = \frac{1}{2} \left(y_i^2 + \mu z_{1,i}^2 + \frac{a_5}{c_1 c_2} z_{2,i}^2 \right) \tag{13.10}$$

with constant $\mu > 0$. Then

$$\dot{V}(y_i, z_{1,i}, z_{2,i}) = y_i u_i - H(y_i, z_{1,i}, z_{2,i})$$

where constants $\gamma_1, \gamma_2 \in (0,1)$, $0 < \mu < \frac{4a_1(1-\gamma_1)\gamma_2}{b_2^2}$ and

$$
\begin{aligned}
H(y_i, z_{1,i}, z_{2,i}) =\, & a_1\gamma_1 y_i^4 - \left(a_2 + \frac{(a_4 + b_3\mu)^2}{4b_4(1-\gamma_2)}\right)y_i^2 - (a_3 + v)y_i \\
& + \left(\mu\gamma_2 - \frac{\mu^2 b_2^2}{4a_1(1-\gamma_1)}\right)z_{1,i}^2 + \mu b_1 z_{1,i} \\
& + \frac{a_5}{c_2}z_{2,i}^2 - a_5 c_3 z_{2,i} \\
& + a_1(1-\gamma_1)\left(y_i^2 + \frac{\mu b_2}{2a_1(1-\gamma_1)z_{1,i}}\right)^2 \\
& + \mu(1-\gamma_2)b_4\left(z_{1,i} - \frac{a_4 - \mu b_3}{2b_4(1-\gamma_2)}y_i\right)^2.
\end{aligned}
$$

Clearly, $H(\cdot) > 0$ for large $\|\mathrm{col}(y_i, z_{1,i}, z_{2,i})\|$, i.e. each Hindmarsh-Rose oscillator satisfies the semipassivity condition (13.2). □

Lemma 13.3. *The system*

$$
\begin{cases}
\dot{z}_{1,i} = -b_1 - b_2 y_i^2 - b_3 y_i - b_4 z_{1,i} \\
\dot{z}_{2,i} = c_1(c_2(y_i + c_3) - z_{2,i})
\end{cases}
\tag{13.11}
$$

is convergent.

Proof. Set $P = I_2$, then the matrix Q as defined in (13.4) is given by

$$
Q = \begin{bmatrix} -b_4 & 0 \\ 0 & -c_1 \end{bmatrix}.
$$

Since $b_4, c_1 > 0$, it follows directly that the condition of Lemma 13.1 is satisfied, i.e. the system (13.11) is convergent. □

Lemmas 13.2 and 13.3 show that assumptions A1 and A2 of Theorem 13.1 hold and therefore we ensure that for sufficiently strong coupling the Hindmarsh-Rose systems (13.9) in the network will (partially) synchronize.

13.4 Experimental setup

We have realized four analog electronic equivalents of the Hindmarsh-Rose equations (13.9), partially based on the implementation as presented in [Lee *et al.*, 2004], with the following set of nominal parameters:

$$
\begin{aligned}
& a_1 = 1, && a_2 = 3, \quad a_3 = 2, && a_4 = 5,\ a_5 = 1, \\
& v = 3.25, && b_1 = 0.8,\ b_2 = 1, && b_3 = 2,\ b_4 = 1, \\
& c_1 = 0.005, && c_2 = 4, \quad c_3 = 2.618. &&
\end{aligned}
$$

Each electronic Hindmarsh-Rose system consists of three integrator circuits, which integrate the Hindmarsh-Rose equations, and a multiplier circuit, build using *AD633j* voltage multipliers, that generates the squared and cubic terms in the Hindmarsh-Rose equations. Figure 13.1 shows a single electronic Hindmarsh-Rose circuit, and Figure 13.2 shows measured chaotic signals from such an electronic oscillator. In particular, in this figure the time series of the states y, z_1, z_2 are depicted and (a part of) the chaotic attractor in the phase-space is shown.

Fig. 13.1   The Hindmarsh-Rose electronic circuit.

There are slight differences between the measured signals and the signals that can obtained through numerical integration of the equations (13.9). This mismatch is due to tolerances of the used components, i.e. the parameters of each circuit differ a little from the nominal ones. This implies that there are small differences between the individual circuits as well. Hence, synchronization in the sense that $x_i = x_j$, where $x_i = [y_i, z_{1,i}, z_{2,i}]^\top$, is not possible. Therefore, we can only expect that the systems will *practically (partially) synchronize* (as defined in Definition 13.1). Hence, we say that two systems i and j practically synchronize whenever $\limsup_{t\to\infty} \|x_i(t) - x_j(t)\| \leq 0.5$ [V].

Remark 13.1. Although the value $\delta = 0.5$ [V] seems rather high, one has to realize that due to the spiking behavior of the signals (see Figure 13.2), a small mismatch induces a relatively large error.

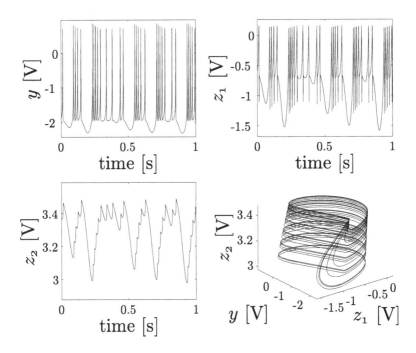

Fig. 13.2   Measured states of the electronic circuit.

In order to define the connections between the Hindmarsh-Rose electronic systems, a synchronization interface is developed that makes use of a microcontroller in which the coupling functions (13.7) can be programmed. The use of a microcontroller to define the coupling functions allows relatively easy experimenting with different network topologies and changes in coupling strength.

13.5   Synchronization experiments

Two systems

Before we discuss synchronization in a network with all four systems, we first consider the case of two diffusively coupled systems being interconnected with coupling strength K.

The two systems are connected by feedback (13.7) with Γ defined as:

$$\Gamma = \begin{bmatrix} K & -K \\ -K & K \end{bmatrix}.$$

It turns out that the two electronic Hindmarsh-Rose oscillators practically synchronize when $K \geq 0.6$. This experimentally obtained coupling strength is pretty close to the value $K \geq 0.5$ that is found in simulations. Figure 13.3 shows the practical synchronization of the two systems for $K = 0.6$. In the left pane the y-states of the two systems are shown as function of time. The synchronization phase portrait is depicted on the right pane. The same is shown in Figure 13.4 in case that the coupling between the two systems is twice that large. One can see that the error between both signals decreases when K increases, i.e. δ becomes smaller.

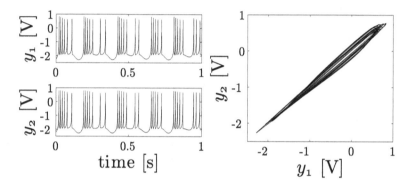

Fig. 13.3   Synchronization of the two coupled systems with $K = 0.6$.

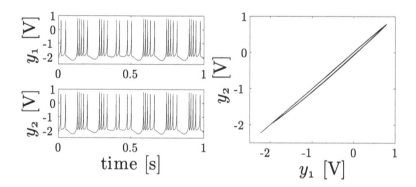

Fig. 13.4   Synchronization of the two coupled systems with $K = 1.2$.

Three systems

Next, three electronic Hindmarsh-Rose oscillators coupled in a ring are considered. The corresponding coupling matrix is given by:

$$\Gamma' = \begin{bmatrix} 2K & -K & -K \\ -K & 2K & -K \\ -K & -K & 2K \end{bmatrix}.$$

Using a conjecture stated by Wu and Chua (1996), the coupling strength required to synchronize the three systems can be determined from the coupling that is required to synchronize two systems. The conjecture is formulated as follows: given two networks of diffusively coupled systems, if the systems in the network with coupling matrix Γ synchronize, then the systems in the network with coupling matrix Γ' synchronize if and only if $\gamma = \gamma'$, where γ and γ' denote the smallest nonzero eigenvalues of Γ and Γ', respectively. Applying the Wu-Chua conjecture we expect the systems to synchronize when $K \geq 0.4$ in the experimental setup and for $K \geq 0.34$ in the simulation. Indeed, the three connected Hindmarsh-Rose oscillators show synchronized behavior for $K = 0.34$ in a simulation study, and the three experimental systems in the network practically synchronize when $K \geq 0.4$, see Figure 13.5.

Remark 13.2. Although it can be shown that the Wu-Chua conjecture is not valid in general, cf. [Pecora, 1998], it is stated in [Pogromsky and Nijmeijer, 2001] that the conjecture is true for systems satisfying assumption A2 of Theorem 13.1.

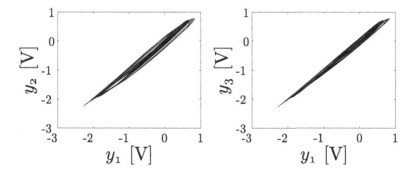

Fig. 13.5 Synchronization phase portrait for three systems ($K = 0.4$).

Four systems

The four systems are coupled in a ring as shown schematically in Figure 13.6.

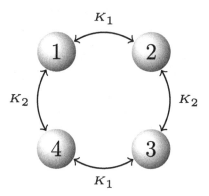

Fig. 13.6   Setup with four systems.

The corresponding coupling matrix for this setup is given as

$$\Gamma'' = \begin{bmatrix} K_1 + K_2 & -K_1 & 0 & -K_2 \\ -K_1 & K_1 + K_2 & -K_2 & 0 \\ 0 & -K_2 & K_1 + K_2 & -K_1 \\ -K_2 & 0 & -K_1 & K_1 + K_2 \end{bmatrix}.$$

The network described by the matrix Γ'' does possess some symmetries. The following matrices define a permutation of the network:

$$\Pi_1 = \begin{bmatrix} E & O \\ O & E \end{bmatrix}, \quad \Pi_2 = \begin{bmatrix} O & E \\ E & O \end{bmatrix}, \quad \Pi_3 = \begin{bmatrix} O & I_2 \\ I_2 & O \end{bmatrix},$$

and $\Pi_4 = I_4$, where O denotes the 2×2 matrix with all its elements equal to zero and

$$E = \begin{bmatrix} 0 & 1 \\ 1 & 0 \end{bmatrix}.$$

The action of Π_1 is to switch the systems 1 and 2, and simultaneously switching of the systems 3 and 4. It follows immediately from Figure 13.6 that the network is left invariant with respect to its interconnections. The matrices Π_2 and Π_3 define similar actions, while Π_4 leaves everything unchanged.

The matrices Π_1, Π_2 and Π_3 define, respectively, the following linear invariant manifolds:

$$\mathcal{A}_1 = \left\{ x \in \mathbb{R}^{12} | x_1 = x_2, x_3 = x_4 \right\},$$
$$\mathcal{A}_2 = \left\{ x \in \mathbb{R}^{12} | x_1 = x_4, x_2 = x_3 \right\},$$
$$\mathcal{A}_3 = \left\{ x \in \mathbb{R}^{12} | x_1 = x_3, x_2 = x_4 \right\}.$$

Applying Theorem 13.1, we have $\lambda' = 2K_1$ for Π_1, $\lambda' = 2K_2$ for Π_2, and $\lambda' = \min(2K_1, 2K_2)$ for Π_3. This means that for large enough K_1 we can expect asymptotic stability of a subset of $\mathcal{A}_1$ and for large enough K_2 a subset of $\mathcal{A}_2$ is asymptotically stable. A subset of $\mathcal{A}_3$ can only be stable as the stable intersection of $\mathcal{A}_1$ and $\mathcal{A}_2$, which describes the fully synchronized state. In our experimental setup practical partial synchronization with respect to the manifold $\mathcal{A}_1$ is found for $K_1 \geq 0.6 > K_2$, while practical partial synchronization with respect to $\mathcal{A}_2$ follows, obviously, when $K_2 \geq 0.6 > K_1$. The phase portraits corresponding to these synchronization regimes are depicted in Figures 13.7 and 13.8, respectively.

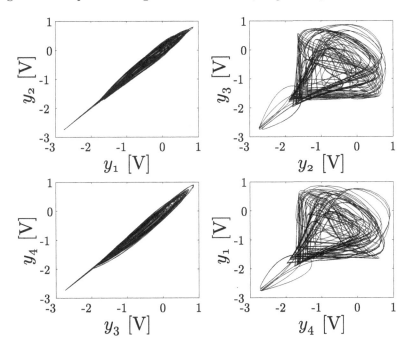

Fig. 13.7   Partial practical synchronization of four systems with respect to the linear invariant manifold $\mathcal{A}_1$ ($K_1 = 0.6$, $K_2 = 0.3$).

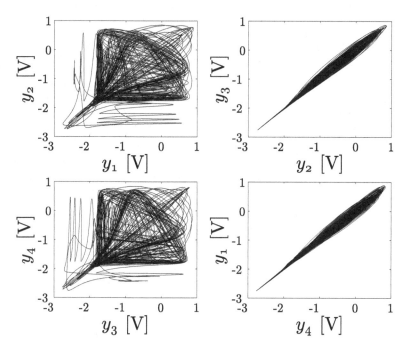

Fig. 13.8   Partial practical synchronization of four systems with respect to the linear invariant manifold $\mathcal{A}_1$ ($K_1 = 0.3$, $K_2 = 0.6$).

Depending on the values of K_1 and K_2, in this ring setup there are two possible routes from no synchronization to full synchronization:

$$\text{no synchrony} \rightarrow \mathcal{A}_1 \rightarrow \mathcal{A}_1 \cap \mathcal{A}_2 \text{ (full synchrony)},$$
$$\text{no synchrony} \rightarrow \mathcal{A}_2 \rightarrow \mathcal{A}_1 \cap \mathcal{A}_2 \text{ (full synchrony)}.$$

The systems in the experimental setup will indeed practically synchronize when $K_1 = K_2 \geq 0.6$. Figure 13.9 shows the synchronization phase portraits for the four systems in case that $K = 0.6$.

13.6   Conclusions

We have presented our experimental finding of synchronous behavior in an experimental setup with up to four coupled electronic, chaotic Hindmarsh-Rose oscillators. At first, it is shown that each free Hindmarsh-Rose system is semipassive and the internal dynamics are convergent. Therefore, under the condition that the coupling between the systems is large enough, the

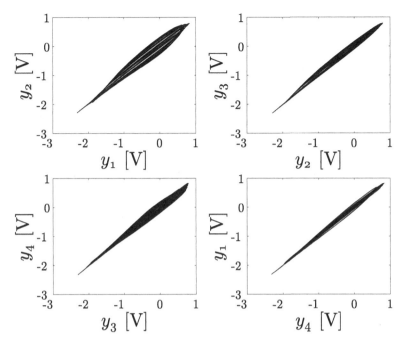

Fig. 13.9   Synchronization phase portraits for four system connected in a ring with the couplings $K_1 = K_2 = 0.6$.

systems in the network should show (partially) synchronized behavior. Indeed, in the experimental setup synchronization (and partial synchronization) in networks consisting of two, three or four oscillators is witnessed. We remark that because of small differences in the behavior of the individual circuits, we are not able to achieve a perfect zero synchronization error, but practical (partial) synchronization is achieved.

Acknowledgments

This research is supported by the Dutch-Russian program on interdisciplinary mathematics *"Dynamics and Control of Hybrid Mechanical Systems"* (NWO grant 047.017.018).

References

Demidovich, B. P. (1967). *Lectures on Stability Theory.* Nauka-Moscow. in Russian.

Gray, C. M. (1994). Synchronous oscillations in neuronal systems: Mechanisms and functions. *J. Comp. Neuroscience* **1**(1-2), 11–38.

Hindmarsh, J. L. and R. M. Rose (1984). A model of neuronal bursting using three coupled first order differential equations. *Proc. R. Soc. London B* **221**(1222), 87–102.

Huijberts, H. J. C., H. Nijmeijer and R. M. A. Willems (1998). A control perspective on communication using chaotic systems. *Proc. 37th IEEE Conf. Decision Contr.* **2**, 1957–1962.

Lee, Y. J., J. Lee, Y. B. Kim, J. Ayers, A. Volkovskii, A. Selverston, H. Abarbanel and M. Rabinovich (2004). Low power real time electronic neuron vlsi design using subthreshold technique. *Proc. IEEE Int. Symp. Circ. and Syst.* **4**, 744–747.

Oud, W. T. and I. Tyukin (2004). Sufficient conditions for synchronization in an ensemble of hindmarsh and rose neurons: Passivity-based approach. *6*[th] *IFAC Symp. Nonlinear Contr. Systems, Stuttgart.*

Pavlov, A. V., N. v. d. Wouw and H. Nijmeijer (2006). *Uniform Output Regulation of Nonlinear Systems.* Birkhäuser Berlin.

Pecora, L. M. (1998). Synchronization conditions and desynchronizing patterns in coupled limit-cycle and chaotic systems. *Phys. Rev. E* **58**(1), 347–360.

Pecora, L. M. and T. L. Carroll (1990). Synchronization in chaotic systems. *Phys. Rev. Lett.* **64**(8), 821–824.

Pikovsky, A., M. Rosenblum and J. Kurths (2003). *Synchronization.* 2 ed.. Cambridge University Press.

Pogromsky, A. Yu. (1998). Passivity based design of synchronizing systems. *Int. J. Bif. Chaos* **8**(2), 295 – 319.

Pogromsky, A. Yu. and H. Nijmeijer (2001). Cooperative oscillatory behavior of mutually coupled dynamical systems. *IEEE Trans. Circ. Syst. I* **48**(2), 152–162.

Pogromsky, A. Yu., G. Santoboni and H. Nijmeijer (2002). Partial synchronization: from symmetry towards stability. *Physica D* **172**(1-4), 65–87.

Rodriguez-Angeles, A. and H. Nijmeijer (2001). Coordination of two robot manipulators based on position measurements only. *Int. J. Contr.* **74**(13), 1311–1323(13).

Strogatz, S. H. and I. Stewart (1993). Coupled oscillators and biological synchronization. *Sci. Am.* **269**(6), 102–109.

Wu, C. W. and L. O. Chua (1996). On a conjecture regarding the synchronization in an array of linearly coupled dynamical systems. *IEEE Trans. Circ. Syst. I* **43**(2), 161–165.

Multipendulum Mechatronic Setup for Studying Control and Synchronization

A.L. Fradkov*, B.R. Andrievskiy*, K.B. Boykov†, B.P. Lavrov*

*Institute for Problems of Mechanical Engineering, RAS,
St. Petersburg, 61, VO, Bolshoy,
Russia

†Corporation Granit, St. Petersburg, Russia

Abstract

In the paper a novel multipendulum mechatronic setup is described. It allows to implement different algorithms of estimation, synchronization and control. The set-up is aimed at solving various research and educational tasks in the areas of hybrid modeling, analysis, identification and control of mechanical systems.

14.1 Introduction

Problems of oscillatory mechanical systems control and synchronization have significant theoretical interest and practical value. There are many papers where this problem was considered and significant results have been achieved [Christini *et al.*, 1996; Blekhman *et al.*, 1997; Andrievsky *et al.*, 1998; Andrievskii *et al.*, 1996; Åström and Furuta, 2000; Andrievsky and Boykov, 2001; Blekhman and Fradkov, 2001; Kumon *et al.*, 2002; Santoboni *et al.*, 2003; Fradkov *et al.*, 2005; Fradkov, 2005, 2007]. For the purposes of research and control engineering education it is important to build up

211

appropriate laboratory equipment and software to work for investigation of this kind of system. In the last decades various mechatronic laboratory setups have been described in the literature: inverted pendulum [Furuta and Yamakita, 1991], Schmid pendulum (reaction-wheel pendulum) [Schmid, 1999; Spong *et al.*, 2001; Andrievsky, 2004], cart-pendulum [Gromov and Raisch, 2003; Gawthrop and McGookin, 2004; Graichen *et al.*, 2007], Furuta pendulum [Furuta *et al.*, 1994; Åström and Furuta, 2000; Suzuki *et al.*, 2004a,b], coupled two-pendulum systems [Andrievsky and Boykov, 2001; Kumon *et al.*, 2002; Fradkov *et al.*, 2002, 2005; Yagasaki, 2007], pendubot [Spong and Block, 1996], humanoid robots [Kim and Oh, 2004], crane systems [Masoud *et al.*, 2004; Wollherr and Buss, 2003], pendulum-like juggling system [Suzuki *et al.*, 2003], force feedback paddle [Saigo *et al.*, 2003], laboratory model helicopters [Apkarian, 1999; Andrievsky *et al.*, 2007] etc. The usage of such equipment is threefold. Firstly, such units are useful for research since they may serve as testbeds for testing new control algorithms under real world constraints. Secondly, it is used for education, allowing students to enhance their skills in control systems design. Finally, it may be used for demonstration, attracting newcomers to the control systems area.

In the present paper the multipendulum mechatronic setup, designed in the Institute for Problems of Mechanical Engineering of RAS (Saint Petersburg, Russia) in the framework of the Russia–Netherlands cooperation program.

The multipendulum mechatronic setup of IPME RAS includes:
– a modular multi-section mechanical oscillating system;
– an electrical equipment (with computer interface facilities);
– the personal computer for experimental data processing, representation of the results the real-time control.

For data exchange via standard In-Out ports of the computer, the special exchange routine is written. The devices are connected by means of the elastic link. In Sec. 14.2 a brief description of the construction is presented. For making laboratory experiments and on-line control, electrical design, data exchange interface and software tools were created. Their description is given in Sec. 14.3.

14.2   Design of mechanical part

The setup consists of a number of identical pendulum sections connected with springs. The schematics of a pendulum section is presented in Fig. 14.1, while a photo of four sections is given in Fig. 14.2. The foundation of the section is a hollow rectangular body. Inside the body an electrical magnet and electronic controller board are mounted. On the foundation the figure support containing the platform for placing the sensors in its middle part is mounted. The pendulum itself possesses a permanent magnet tip in the bottom part. The working ends of the permanent magnet and the electrical magnet are posed exactly opposite each other and separated with a non-magnetic plate in a window of the body. The idea behind control of the pendulum is changing the poles of the electrical magnet by means of switching the direction of the current in the windings of the electrical magnet. In order to allow changing the eigenfrequency of the pendulum oscillations the pendulum is endowed with additional plummets and counterparts changing its effective length (the distance between the suspension point and the center of mass. On the rotation axis of the pendulum the optical encoder disk for measuring the angle (phase) of the pendulum is mounted. It has 90 slits. The peripheral part of the disk is posed into the slit of the sensor support. The sensor consists of a radiator (emitting diode) and a receiver (photodiode). The obtained sequences of signals allow to measure angle (phase) and angular velocity of the pendulum, evaluate amplitude and crossing times and other variables related to the pendulum dynamics.

Axes of the neighbor sections are connected with the torsion springs, arranging force interaction and allowing exchanging energy between neighbor sections. The set of interconnected pendulum sections represents a complex oscillatory dynamical system, characterized by nonlinearity and high number of degrees of freedom. Such a mechanical system can serve as a basis for numerous educational and research experiments related to dynamics, control and synchronization in the networks of multidimensional nonlinear dynamical systems. In principle, any number of sections can be connected. At the moment mechanical parts of 50 sections are manufactured.

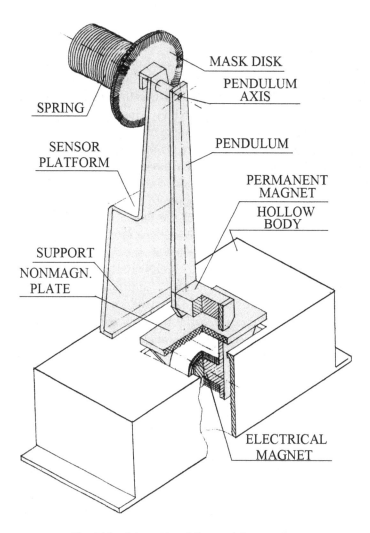

Fig. 14.1   Schematics of the pendulum section.

14.3   Electronics of the multipendulum setup

Oscillation control is provided on the basis of combined hardware/software
method. The energy for excitation is transmitted by the pulse-width modulated (PWM) signal with the constant level and variable duty cycle. From
the programming point of view, hardware is represented by the write-only
registers (WO) for putting in the prescribed duty cycle of control signal

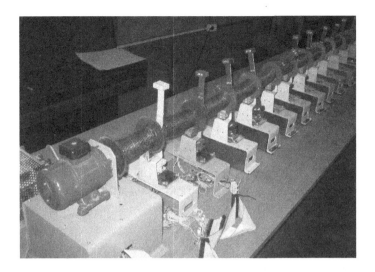

Fig. 14.2   Photo of 12 coupled pendulum sections.

from the computer, and the read-only registers (RO) for transfer oscillation halfperiod duration values to the computer. The PWM based method provides more precise control than the number-pulse one, because of averaging the high frequency pulses by the mechanical subsystem. The *control unit* generates the exciting action applied to the pendulums via the opposite magnetic fields. It includes bi-channel *asynchronous pulse-width modulator* (APWM), the Data Exchange System and the power amplifiers to drive the electromagnets.

14.3.1   *System for data exchange with control computer*

The Data Exchange System of the setup is intended to transfer data and control commands from the Control Computer to the interface board of the pendulum sections. Each interface board is an intelligent measuring/controlling electronic devise, assigned for unloading processor of the Control Computer from chores of forming the control signal and preventing the Control Computer from a wasteful wait state of the sensor replies.

A main problem for designing the Data Exchange System was a demand to increase the channel capacity, preventing, at the same time, any loss of synchronism in interaction between the Control Computer and the pendulum sections. Since the setup consists of a large number (up to 50) pendulum sections, and the number of enabled sections may be different for different experiments, and also the sections arrangement may be modified, designing of data exchange interface is a hard problem. Demand of hardware independence and universality of the data exchange interface should be also taken into account since it makes possible ensuring compatibility of the stand with standard PCs and the microcontroller-based information processing systems as well.

14.3.2 *Architecture of the Data Exchange System*

The most reasonable architecture for the Information Management Systems, including $5 - 8$ assemblies and more is a *data-bus* one. The bus of the multipendulum setup works in a *bidirectional* mode. It consists of:

- the bidirectional data link;
- the control lines;
- the confirmation line;
- the power supply lines.

Hardwired noise immunity is ensured by terminating resistors insertion to the both end of the bus. To prevent the bus-conflict accidents, caused by faults of the setup units, the open collector (open-drain) transmitters and inverting receivers are exploited in the data-bus design. Using this technology makes it possible to restrict an abnormal current for each single bus line choosing resistance of the terminating resistor. A total length of the data bus is 14 m.

14.3.3 *Computer-process interface*

Different types of interface, applicable both for PCs and special-purpose hardware, such as a serial port (USART and USB); parallel port (SPP, ECP, EPP), the Fast Ethernet channel, IEEE1394 interface, and the IDE channel were analyzed. The main requirements to the interface are as follows:

- ability of the direct access to registers;
- byte-wide or word-wide mode of operation;
- high carrying capacity (no less than 1 MB/s);
- data-bus architecture capability.

The EPP (Enhanced Parallel Port) was chosen as the most efficient interface, allowing to organize synchronous data exchange and a data-bus architecture of the communication link. For matching the data bus and the standard parallel port of the PC, an add-on device, the *dispatcher* was designed. The dispatcher is intended for:

- inversion of received and transmitted data;
- front-side bus signal generation;
- generation of waiting intervals in the absence of the bus response;
- generation of the "error" reply in the absence of the bus response during the waiting interval;
- processing of the EPP data communications protocol for the data exchange with PC, complying with the IEEE1284 standard.

14.3.4 *Electronic modules of the set-up*

Pendulum sections and electric motors are equipped with the electronic modules. The electronic modules have following functions:

- data exchange with the data bus;
- generation of control signals for executive devices (the pendulum actuating coils and the electric motors);
- processing of the sensor signals

Control of pendulums is performed by pulse-width modulation (PWM) of the electric voltage applied to the actuating coil of the electric magnet mounted in the basement of the pendulum section (see Fig. 14.1). The magnetic field strength is a rapidly decreasing function of a distance between the electric magnet and the pendulum bob. Therefore in the self-directed mode the control voltages are automatically applied to the actuating coil when the pendulum bob is in the vicinity of its lower position. The measured quantities of each pendulum section are the *angular displacement* and the *driving direction* of the bob. The angular encoder, mounted in alignment with the pendulum axis, is used as a sensor. Pulse patterns from the encoder outputs are transformed by the logical units of the electronic modules into binary codes, which may be read out under data bus requests.

Control of the electric motors is implemented by applying the pulse-width modulated voltage to the armature coil. For changing the direction of the motor driving torque, the double-channel PW modulators and the bridge power amplification circuits are used in the output stages of motor control.

The electronic modules of the dispatcher, of the pendulum sections and motors control are designed based on single-type solutions employing the *All-in-One Logical Cards* (AOLCs). The kernel component of AOLC is Field Programmable Gate Array chip EPM240T100C5 with the logical capacity in 240 macrocells, manufactured by *Altera* Co. Logic synthesis was made with the help of *Quartus II Design Software*, ver 5.1. The chip resources are used up to 80% of the full capacity.

The machine code of the electronic modules includes: write instruction for controlling parameters; read instruction for measured data; executive instruction for switching the module mode (self-acting or software-based modes).

14.3.5 *Communications protocol*

The communications protocol secures writing the instructions and command qualifiers to the interface board of the pendulum sections and reading the data, measured by the interface board sensors. The communications protocol uses three kinds of passing: address passing, instruction (mode) passing and data passing. The communications protocol is based on explicit addressing with use of Enhanced Parallel Port (EPP) registers of the PC in accordance with the IEEE 1284 Parallel Port Standard: address register = (base address+3), data register = (base address+4). The registers are available for read-write operations.

For transfer modes ensuring the address register of the EPP-port is used. For logical separation of the data and instruction flows is made by means of two high stages of the address register. The rest five stages of this register contain the module address for access on the next R/W cycle.

Described data exchange protocol ensures exchange rate up to 9615 Hz for simultaneous operation over the bus with all 52 modules and up to 500 kHz for operation with a single module (delays for data processing in PC are not taken into account in the above estimate).

14.4 Conclusions

The system is currently being tested and tuned. Five active and up to 46 passive sections are ready for connection. Already at this stage the system can be used for demonstration and for research. Connecting five pendulum sections one can demonstrate either inphase and antiphase synchroniza-

Fig. 14.3   Inphase synchronization in the chain of five pendulums.

tion in the chain of pendulums, excited by the external harmonic torque, see Figs. 14.3, 14.4. The experiments show that the type of the synchronization mode (inphase or antiphase) depends on the excitation frequency. Namely, there exist certain frequencies ω_1 and ω_2 such that motion of the pendulums close to inphase if the excitation frequency ω less than ω_1 and is approximately antiphase if $\omega > \omega_2$. It was also demonstrated that the pendulum chain may play a role of a mechanical band pass filter with the bandwidth $[\omega_1, \omega_2]$. Inside this interval a kind of traveling wave appears in the chain. These results agree with the conclusions of [Fradkov and Andrievsky, 2007; Andrievsky and Fradkov, 2009].

Some new features of the set-up were discovered during first experiments. For example it appears that it is easy to change the stiffness characteristics of the spring by means of changing distance between the sections. Increasing spring stiffness allows expanding the regions of attraction of the stable synchronous mode.

Fig. 14.4   Antiphase synchronization of in the chain of five pendulums.

Acknowledgments

The work was supported by the Russian Foundation for Basic Research (projects RFBR 08-01-00775, 09-08-00803), the Russia–Netherlands cooperation program (project NWO–RFBR 047.011.2004.004), the Council for grants of the RF President to support young Russian researchers and leading scientific schools (project NSh-2387.2008.1), and the Program of basic research of OEMPPU RAS #2 "Control and safety in energy and technical systems".

References

Andrievsky, B. R. (2004). Stabilization of the inverted reaction wheel pendulum, in A. Fradkov (ed.), *Control in Physical and Technical Systems* (Nauka, St. Petersburg), pp. 52–71, (in Russian).
Andrievsky, B. R. and Boykov, K. B. (2001). Numerical and laboratory experiments with controlled coupled pendulums, in *Prepr. Nonlinear Control Design Symposium NOLCOS'01* (St. Petersburg, Russia), pp. 824–829.

Andrievsky, B. R., Fradkov, A. L., Konoplev, V. A. and Konjukhov, A. P. (1998). Control, state estimation and laboratory experiments with oscillatory mechanical system, in *Prepr. Nonlinear Control Design Symposium NOLCOS98* (Twente, Holland), pp. 761–764.

Andrievsky, B. and Fradkov, A. L. (2009) Behavior analysis of harmonically forced chain of pendulums, in *3rd IEEE Multi-Conference on Systems and Control, MSC'2009* (Saint Petersburg, Russia), pp. 1563–1567.

Andrievskii, B. R., Guzenko, P. Yu. and Fradkov, A. L. (1996) Control of nonlinear vibrations of mechanical systems via the method of Velocity Gradient, *Automation and Remote Control* **57**, 4, pp. 456–467.

Andrievsky, B., Peaucelle, D. and Fradkov, A. L. (2007). Adaptive control of 3DOF motion for LAAS Helicopter Benchmark: Design and experiments, in *Proc. 2007 Amer. Control Conf* (New York, USA), pp. 3312–3317.

Apkarian, J. (1999). Internet control, Circuit Cellar, http://www.circuitcellar.com.

Åström, K. and Furuta, K. (2000). Swinging up a pendulum by energy control, *Automatica* **36**, 2, pp. 287–295.

Blekhman, I. I. and Fradkov, A. L. (eds.) (2001). *Control of Mechatronic Vibrational Units* (Nauka, St.Petersburg), in Russian.

Blekhman, I. I., Fradkov, A. L., Nijmeijer, H. and Pogromsky, A. Y. (1997). On self-synchronization and controlled synchronization of dynamical systems, *Systems & Control Letters* **31**, 5, pp. 299–306.

Christini, D. J., Collins, J. J. and Linsay, P. S. (1996). Experimental control of highdimensional chaos: The driven double pendulum, *Phys. Rev. E* **54**, 5.

Fradkov, A., Andrievsky, B. R. and Boykov, K. B. (2002). Numerical and experimental excitability analysis of multi-pendulum mechatronics system, in *Proc. 15th IFAC World Congress* (Barcelona, Spain).

Fradkov, A., Andrievsky, B. R. and Boykov, K. B. (2005). Control of the coupled double pendulums system, *Mechatronics* **15**, pp. 1289–1303.

Fradkov, A. L. (2005). Application of cybernetical methods in physics, *Physics–Uspekhi* **48**, 2, pp. 103–127.

Fradkov, A. L. (2007). *Cybernetical Physics: From Control of Chaos to Quantum Control* (Springer-Verlag, Berlin-Heiderlberg).

Fradkov, A. L. and Andrievsky, B. (2007). Synchronization and phase relations in the motion of two-pendulum system, *Int. J. Non-Linear Mechanics* **42**, 6, pp. 895–901.

Furuta, K. and Yamakita, M. (1991). Swing up control of inverted pendulum, in *Proc. IECON91*, pp. 2193–2198.

Furuta, K., Yamakita, M., Kobayasji, S. and Nishimura, M. (1994). A new inverted pendulum apparatus for education, in *Proc. IFAC Symp. on Advances Contr. Education* (Tokyo), pp. 191–194.

Gawthrop, P. J. and McGookin, E. (2004). A LEGO-based control experiment, **24**, 5, pp. 43–56.

Graichen, K., Treuer, M. and Zeitz, M. (2007). Swing-up of the double pendulum on a cart by feedforward and feedback control with experimental validation, *Automatica* **43**, 1, pp. 63–71.

Gromov, D. and Raisch, J. (2003). Hybrid control of a cart-pendulum system with restrictions on the travel, in *Proc. Int. Conf. Physics and Control*, Vol. 4 (St. Petersburg, Russia), pp. 1231–1235.

Kim, J. H. and Oh, J. H. (2004). Realization of dynamic walking for the humanoid robot platform KHR-1, *Adv. Robot.* **18**, 7, pp. 749–768.

Kumon, M., Washizaki, R., Sato, J., nd I. Mizumoto, R. K. and Iwai, Z. (2002). Controlled synchronization of two 1-DOF coupled oscillators, in *Proc. 15th Triennial World Congress of IFAC* (Barcelona, Spain).

Masoud, Z. N., Nayfeh, A. H. and Mook, D. T. (2004). Cargo pendulation reduction of ship-mounted cranes, *Nonlinear Dyn.* **35**, 3, pp. 299–311.

Saigo, M., Tani, K. and Usui, H. (2003). Vibration control of a traveling suspended system using wave absorbing control, *Vib. Acoust.-Trans. ASME* **125**, 3, pp. 343–350.

Santoboni, G., Pogromsky, A. Y. and Nijmeijer, H. (2003). Application of partial observability for analysis and design of synchronized systems, *Chaos, Solitons and Fractals* **13**, 1, pp. 356–363.

Schmid, C. (1999). An autonomous self-rising pendulum. Invited paper, in *Proc. European Control Conference ECC'99* (Karlsruhe).

Spong, M. W. and Block, D. (1996). The Pendubot: A mechatronic systems for control research and education, in *Proc. 35th IEEE Conf. Dec. Control (CDC'96)* (New Orleans, USA), pp. 555–556.

Spong, M. W., Corke, P. and Lozano, R. (2001). Nonlinear control of the Reaction Wheel Pendulum, *Automatica* **37**, pp. 1845–1851.

Suzuki, S., Furuta, K. and Shiratori, S. (2003). Adaptive impact shot control by pendulum-like juggling system, *Int. J. Ser. C-Mech. Syst. Mach. Elem. Manuf.* **46**, 3, pp. 973–981.

Suzuki, S., Furuta, K., Sugiki, A. and Hatakeyama, S. (2004). Nonlinear optimal internal forces control and application to swing-up and stabilization of pendulum, *Dyn. Syst. Meas. Control-Trans. ASME* **126**, 3, pp. 568–573.

Suzuki, S., Pan, Y., Furuta, K. and Hatakeyama, S. (2004). VS-control with time-varying sliding sector – Design and application to pendulum, *Asian J. Control* **6**, 3, pp. 307–316.

Wollherr, D. and Buss, M. (2003). Cost-oriented virtual reality and real-time control system architecture Robotica, **21**, 3, pp. 289–294.

Yagasaki, K. (2007). Extension of a chaos control method to unstable trajectories on infinite - or finite-time intervals: Experimental verification, *Phys. Lett. A* **368**, 3,4, pp. 222–226.

High-frequency Effects in 1D Spring-mass Systems with Strongly Non-linear Inclusions

B.S. Lazarov, S.O. Snaeland, J.J. Thomsen

Department of Mechanical Engineering
Technical University of Denmark
Denmark

Abstract

This work generalises the possibilities to change the effective material or structural properties for low frequency (LF) wave propagation, by using high-frequency (HF) external excitation combined with strong non-linear and non-local material behaviour. The effects are demonstrated on 1D chain-like systems with embedded non-linear parts, where the masses interact with a limited set of neighbour masses. The presented analytical and numerical results show that the effective properties for LF wave propagation can be altered by establishing HF standing waves in the non-linear regions of the chain. The changes affect the effective stiffness and damping of the system.

15.1   Introduction

Changing the wave propagation speed by using high-frequency effects in 1D discrete systems with non-linear inclusions and particle interaction forces between the closest neighbours only has been demonstrated in [Lazarov et al., 2008]. The study has been advanced further in more details and numerically verified in [Lazarov and Thomsen, 2009]. In many cases the

223

particle forces are non-local, i.e., they depend on the relative distance or velocities between a particle and a set of neighbour particles. An example of such a system is a chain of magnets oriented in such a way, that the force between two neighbours is repulsive. The possibilities to change the effective material or structural properties for low frequency (LF) wave propagation, by using high-frequency (HF) external excitation combined with strong non-linear and non-local material behaviour, are further generalised in this work. Dynamical changes in materials and structures can be utilised in many problems in structural and mechanical engineering, e.g., in creating elasto-mechanical filters, waveguides, ultrasonic devices, opto-mechanical components, sound and vibration isolation, and in health monitoring.

General HF effects in non-linear systems have been studied analytically, numerically and experimentally in many scientific works, e.g., [Blekhman, 2000; Thomsen, 2003; Thomsen, 2002; Thomsen, 1999; Tcherniak and Thomsen, 1998; Stephenson, 1908; Jensen, 2000; Fidlin, 2000; Fidlin, 2005; Chatterjee et al., 2003; Chatterjee et al., 2004]. HF excitation can change various aspects of the system behaviour. This work utilises the so-called stiffening effect [Thomsen, 2002] and the non-linear damping effect [Chatterjee et al., 2003]. The stiffening effect results in changes of the effective stiffness for low frequency excitation and of the natural frequencies of the system, and the non-linear damping effect alters the effective damping for low frequency excitation. HF excitation can also influence the equilibrium states and their stability. A well known example of the stiffening effect is the upside-down equilibrium stabilisation of the Kapitza pendulum on a vibrating support [Kapitza, 1951], which belongs to the class of *parametrically* excited systems. The effects of HF parametric excitation are well understood, and have been studied analytically and numerically in many articles [Jensen, 2000; Thomsen, 1999; Thomsen, 2002]. Experimental verification for various physical systems is presented in [Jensen et al., 2000; Yabuno and Tsumoto, 2007; Thomsen, 2003; Thomsen, 2007]. Stiffening can also be observed in *externally* excited non-linear systems. Analytical and numerical results for calculating the averaged motion of an externally excited system are presented in [Thomsen, 2008]. A major part of the references cited above address single degree of freedom systems. HF effects for a general class of multi degree of freedom (MDOF) systems have been studied in [Blekhman, 2000; Thomsen, 2002], and later for continuous systems in [Thomsen, 2003].

It has been suggested in recent works [Blekhman, 2000; Lurie, 2007;

Thomsen and Blekhman, 2007] to use modulation of the structural/material characteristics in space and time, in order to create materials and structures with adjustable effective properties. Such materials are known as dynamic materials [Blekhman, 2000; Lurie, 2007]. In [Lurie, 2007; Sorokin *et al.*, 2001] HF parametric excitation has been applied to control the effective material properties. In [Thomsen and Blekhman, 2007] by assuming the existence of a HF standing wave in a continuous non-linear systems, the effective material properties for the LF motion (slow motion) are derived analytically. Compared to the approaches in [Lurie, 2007; Sorokin *et al.*, 2001], where distributed actuation and control are used, the advantage of the approach in [Thomsen and Blekhman, 2007] is the use of a single excitation source for generating the HF field. However, in contrast to the parametric case, where the stiffness change depends on the frequency of the excitation, for the externally excited non-linear system it does not depend on the frequency of the excitation, and is proportional to the response amplitude and the strength of the non-linearity. If the non-linearities are cubic, the change in the effective material stiffness is predicted to be proportional to the square of the HF response amplitude. The same effect for undamped Fermi-Pasta-Ulam chains has also been predicted analytically in [Kosevich, 1993].

The stiffening can be amplified by increasing the amplitude of the HF standing wave and/or by increasing the strength of the non-linearity. Similarly the effective damping is proportional to the amplitude multiplied by the excitation frequency and the strength of the non-linearity of the damping function. Establishing stable finite amplitude waves in a non-linear medium is not a trivial matter in the general case. Numerical and analytical studies of non-linear chains [Rodriguez-Achahc and Perez, 1997; Budinsky and Bountis, 1983] show, that the energy level per particle necessary to start chaotic behaviour increases with fewer particles in the chain, and thus stable standing waves are easier to establish in shorter chains. This property, combined with the fact that the waves with frequencies above a given threshold cannot propagate along linear chains [Brillouin, 1953], is used in the present work, in order to create a HF standing wave in a finite non-linear region around the location of the HF excitation. In this way some of the problems related to stability of the HF response can be avoided. Alternative ways for localising the HF response are also discussed.

Various methods have been used in the past, to analyse the HF effects in linear and non-linear systems, e.g. [Blekhman, 2000; Thomsen, 2003; Thomsen, 2002; Tcherniak and Thomsen, 1998; Fidlin, 2005]. Here the

Method of Direct Separation of Motion (MDSM) is utilised. The method originates from the work [Kapitza, 1951]. Later it was generalised by Blekhman [Blekhman, 2000], and is used in [Thomsen, 2003; Thomsen, 2002; Thomsen, 2007; Thomsen and Blekhman, 2007; Thomsen, 2008; Jensen, 2000; Jensen *et al.*, 2000; Fidlin, 2005, Lazarov and Thomsen, 2009]. The main assumption in MDSM is that the motion can be represented as a superposition of fast and slow oscillations. The equations for the slow motion are obtained by averaging.

The HF response localisation have been applied for controlling the wave speed in chains with non-linear inclusions in [Lazarov *et al.*, 2008; Lazarov and Thomsen, 2009]. In [Lazarov and Thomsen, 2009] it has been demonstrated that HF excitation in a metamaterial with essentially non-linear inclusions can make the material transparent for low frequency waves. HF excitation can also create an effective static force by using the localisation effect and strong non-linearities. In this work the considered system is a 1D lattice consisting of particles with equal mass and interaction forces depending on the relative distances and velocities between a particle and a few of its neighbours. Short non-linear inclusions are embedded in longer linear chains, and the HF excitation is explicitly applied in a non-parametric way in the regions with non-linearities. The HF frequency is within the stop band of the linear chain, and the HF response becomes localised around the location of the applied excitation. An example of a 1D lattice with non-linear inclusions and interaction forces only between neighbour particles is shown in Fig. 15.1.

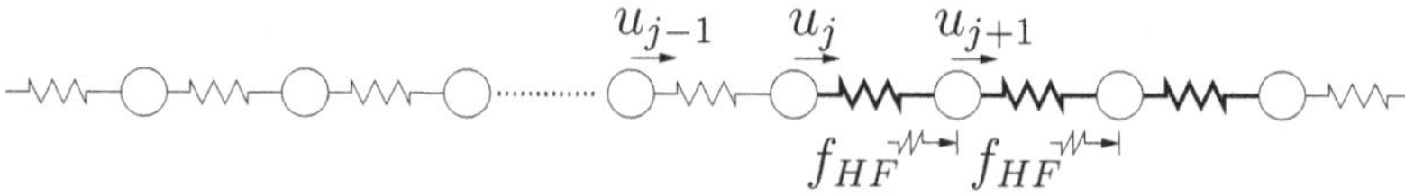

Fig. 15.1   Spring mass system.

The presentation is separated into two major parts - theoretical analysis and numerical testing. As the band gap effect plays and important role in localising the HF response, the wave propagation properties of linear chains with interaction forces between neighbour masses is discussed after introducing the equations of motion. Alternative ways for localising the HF response are also briefly outlined. The MDSM is used to obtain the equations governing the slow and fast motions for the non-linear inclusions and the obtained analytical results and predictions for the stiffening effect are tested later by numerical simulations.

15.2 Mechanical model

The considered system consists of particles with mass m interacting with a limited set of neighbour particles. The oscillators are labelled with an index j, and their displacements around the static equilibrium position are denoted with u_j. For a system with N masses, the equations of motion can be written in a matrix-vector form as

$$\frac{d^2\mathbf{u}}{dt^2} + \mathbf{f}_D\left(\frac{d\mathbf{u}}{dt}\right) + \mathbf{f}_R(\mathbf{u}) = \mathbf{f}(t) \tag{15.1}$$

where $\mathbf{u}(t) \in R^N$ is a vector with an element j equal to the displacements of the jth mass, and $\mathbf{f}_D(\cdot)$ and $\mathbf{f}_R(\cdot)$ are vector functions with the damping and the restoring forces respectively. The damping forces are assumed to depend only on the velocities of the particles in the system, and the restoring forces depend on the relative displacements between the masses. The damping and the restoring forces acting on mass j are modelled by using the following expressions

$$f_{D,j} = \sum_{i=1}^{n_D}\sum_{k=1}^{m_D}\left[a^D_{j,k,i}r^-_{j,i}(\mathbf{v})^k + b^D_{j,k,i}r^+_{j,i}(\mathbf{v})^k\right] \tag{15.2}$$

$$f_{R,j} = \sum_{i=1}^{n_R}\sum_{k=1}^{m_R}\left[a^R_{j,k,i}r^-_{j,i}(\mathbf{u})^k + b^R_{j,k,i}r^+_{j,i}(\mathbf{u})^k\right] \tag{15.3}$$

where

$$r^-_{j,i}(\mathbf{x}) = x_j - x_{j-i} \tag{15.4}$$

$$r^+_{j,i}(\mathbf{x}) = x_j - x_{j+i} \tag{15.5}$$

and $\mathbf{v} = d\mathbf{u}/dt$. The external excitation $\mathbf{f}(t)$ is assumed to consists of LF and HF components:

$$\mathbf{f}(\omega t, \Omega t) = \mathbf{f}_{LF}(\omega t) + \mathbf{f}_{HF}(\omega t, \Omega t) \tag{15.6}$$

where the high frequency Ω is much larger than the low frequency ω. In addition the HF component is assumed to fulfil the following condition:

$$\int_{\tau_0}^{\tau_0+2\pi} \mathbf{f}_{HF}(t_s, \tau)\,d\tau = 0 \tag{15.7}$$

where $\tau = \Omega t$ and $t_s = \omega t$ are called fast and slow time respectively, and t_s is assumed to be independent from τ in the integration.

15.3   Band gap effects in periodic structures

In discrete lattices waves can propagate within specific bands of frequencies known as pass bands, and attenuate within bands of frequencies known as stop bands or band gaps. In order to demonstrate these properties, a simple linear chain with equal masses is considered here. Wave propagation in such lattices, with interaction forces depending only on the relative distance between the neighbours, reveals many of the effects which can be observed in more complex models. The motion of the jth mass is governed by the following equation:

$$\ddot{u}_j + \zeta\omega_0\left(2\dot{u}_j - \dot{u}_{j-1} - \dot{u}_{j+1}\right) + \omega_0^2\left(2u_j - u_{j-1} - u_{j+1}\right) = 0 \qquad (15.8)$$

where overdot denotes differentiation with respect to the time, $\omega_0 = \sqrt{k/m}$, k is the stiffness of the linear spring connecting two neighbour masses, and m is the mass of one particle. The solution of (15.8) is sought in the form

$$u_j = Ae^{j\mu - i\omega t} \qquad (15.9)$$

where μ is the so-called wave propagation constant of harmonic wave with frequency ω, and j is the mass index. The wave propagation constant is equal to the wave number multiplied by the distance between the masses. A relation between the wave frequency and the propagation constant can be obtained by inserting (15.9) into (15.8) in the form:

$$\cosh\left(\mu\right) = 1 + \frac{1}{2}\frac{\left(\omega/\omega_0\right)^2}{i\zeta\left(\omega/\omega_0\right) - 1} \qquad (15.10)$$

Plots of the real and imaginary parts of μ for different values of the damping parameter ζ are shown in Fig. 15.2. For undamped chains, the wave propagation constant is real for frequencies above $\omega \geq 2\omega_0$ and purely imaginary for $2 > \omega/\omega_0 \geq 0$. Waves with frequency $\omega \geq 2\omega_0$ are attenuated along the chain. Applying excitation with such frequency will result in a localised response around the exited mass. Waves with frequency $2\omega_0 \geq \omega \geq 0$ propagate unattenuated along the undamped chain. If damping is introduced in the model, all waves are attenuated. Waves with higher frequencies are damped faster than waves with lower frequencies. Therefore, in damped lattices high frequency excitation applied to a particle along the chain will also result in a localised response.

The group velocity for linear undamped chains is defined as the derivative of the wave frequency with respect to the wave number, and in the normalised case is the derivative of ω with respect to μ. For a wave with frequency ω the group velocity is increased by increasing ω_0. For the damped case increasing the damping results in an increase of the slope of the dispersion curve and, therefore, a pulse with narrow band of frequencies will propagate faster by increasing the parameter ζ.

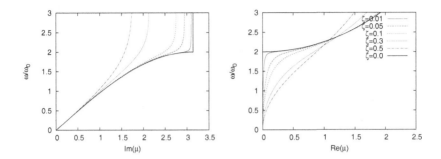

Fig. 15.2   Dispersion relation for different damping ratios.

If the chain has periodic inclusions, the pass band can be split by one or many band gaps. A detailed study can be found in [Brillouin, 1953]. If the restoring force depends on the relative distance between mass j and particles $j \pm k, k > 1$, the dispersion relation is not longer one to one mapping between μ and ω. For a single frequency, multiple wave lengths can be observed. However, similar to the more simple case with interaction between the neighbour masses, band gaps can be observed in the dispersion relations. A similar effect can be obtained in lattices with random inclusions. Using periodic structures allows for utilising the localisation effect also in continuous models. In the case when inclusions are used, the wave lengths in the stop band are comparable with the distance between the inclusions. Localisation in the lower frequency range is also possible by attaching single or multiple linear or non-linear oscillators to the particles [Lazarov and Jensen, 2007].

15.4   Approximate equations governing the slow and the fast motion

As mentioned earlier the focus here is on the low frequency system response, and how the high frequency excitation affects the relevant system properties. Simplified equations governing the slow motion are obtained by using the MDSM [Blekhman, 2000; Fidlin, 2005]. The solution to 15.1 is sought in the following form:

$$\mathbf{u}(t) = \mathbf{z}(t_s) + \boldsymbol{\psi}(t_s, \tau) \tag{15.11}$$

where $\mathbf{z}(t_s)$ is the slow or the averaged motion of the system, and $\boldsymbol{\psi}(t_s, \tau)$ is fast oscillatory response. The HF motion $\boldsymbol{\psi}(t_s, \tau)$ is assumed to be periodic in τ with period 2π, and its fast time average is zero:

$$\langle \boldsymbol{\psi}(t_s, \tau) \rangle_\tau = \frac{1}{2\pi} \int_0^{2\pi} \boldsymbol{\psi}(t_s, \tau) \, \mathrm{d}\tau = 0 \tag{15.12}$$

Applying the fast time averaging operator to (15.11) and using (15.12), results in averaged motion $\langle \mathbf{u}(t) \rangle_\tau$, which is equal to the slow motion function $\mathbf{z}(t_s)$. Given the periodicity of $\boldsymbol{\psi}$ and that (15.12) is fulfilled, the following equalities are fulfilled as well

$$\langle \boldsymbol{\psi}(t_s, \tau)' \rangle = \langle \boldsymbol{\psi}(t_s, \tau)'' \rangle = \left\langle \dot{\boldsymbol{\psi}}(t_s, \tau) \right\rangle = 0 \tag{15.13}$$

where overdot denote differentiation with respect to the slow time t_s, and primes denote differentiation with respect to the fast time τ.

By inserting equations (15.11) and (15.6) into (15.1), and applying the fast time averaging operator, a simplified system of equations governing the slow component of the system motion is obtained as:

$$\omega^2 \ddot{\mathbf{z}} + \langle \mathbf{f}_D(\mathbf{v}) \rangle_\tau + \langle \mathbf{f}_r(\mathbf{z} + \boldsymbol{\psi}) \rangle_\tau = \mathbf{f}_{LF}(t_s) \tag{15.14}$$

The expressions for the averaged damping $\langle \mathbf{f}_D(\mathbf{v}) \rangle_\tau$ and the restoring forces $\langle \mathbf{f}_r(\mathbf{z} + \boldsymbol{\psi}) \rangle_\tau$ in (15.14) are obtained by using (15.2) and (15.3), and by expanding the results on LF and HF velocity and displacement differences:

$$\langle \mathbf{f}_D(\mathbf{v}) \rangle_\tau = \mathbf{f}_D(\mathbf{v}_z) + \langle \mathbf{f}_D(\mathbf{v}_\psi) \rangle_\tau$$
$$+ \sum_{i=1}^{n_D} \sum_{k=1}^{m_D-1} \sum_{p=k+1}^{m_D} \left[c_{j,k,i,p}^D r_{j,i}^-(\mathbf{v}_z)^k + e_{j,k,i,p}^D r_{j,i}^+(\mathbf{v}_z)^k \right] \tag{15.15}$$

$$\langle \mathbf{f}_R(\mathbf{u}) \rangle_\tau = \mathbf{f}_R(\mathbf{z}) + \langle \mathbf{f}_R(\boldsymbol{\psi}) \rangle_\tau$$
$$+ \sum_{i=1}^{n_R} \sum_{k=1}^{m_R-1} \sum_{p=k+1}^{m_R} \left[c_{j,k,i,p}^R r_{j,i}^-(\mathbf{u})^k + e_{j,k,i,p}^R r_{j,i}^+(\mathbf{u})^k \right] \tag{15.16}$$

where

$$c_{j,k,i,p}^R = \binom{p}{k} a_{j,k,i}^R \left\langle r_{j,i}^-(\boldsymbol{\psi})^{p-k} \right\rangle_\tau \tag{15.17}$$

$$e_{j,k,i,p}^R = \binom{p}{k} b_{j,k,i}^R \left\langle r_{j,i}^+(\boldsymbol{\psi})^{p-k} \right\rangle_\tau \tag{15.18}$$

The coefficients $c_{j,k,i,p}^{D}$ and $e_{j,k,i,p}^{D}$ can be obtained by using equations (15.17) and (15.18), replacing ψ with $\mathbf{v}_\psi$ and the upper index R with D. The vectors $\mathbf{v}_z = \mathrm{d}\mathbf{z}/\mathrm{d}t$ and $\mathbf{v}_\psi = \mathrm{d}\psi/\mathrm{d}t$ are the derivatives of $\mathbf{z}$ and ψ with respect to the time t.

Equations (15.15) and (15.16) can be written as

$$\langle \mathbf{f}_D(\mathbf{v}) \rangle_\tau = \mathbf{f}_D^{eq}(\mathbf{v}_z) + \langle \mathbf{f}_D(\mathbf{v}_\psi) \rangle_\tau \tag{15.19}$$

$$\langle \mathbf{f}_R(\mathbf{u}) \rangle_\tau = \mathbf{f}_R^{eq}(\mathbf{z}) + \langle \mathbf{f}_R(\psi) \rangle_\tau \tag{15.20}$$

where

$$\mathbf{f}_D^{eq}(\mathbf{v}_z) = \sum_{i=1}^{n_D} \sum_{k=1}^{m_D} \left[a_{j,k,i}^{D,eq} r_{j,i}^{-}(\mathbf{v}_z)^k + b_{j,k,i}^{D,eq} r_{j,i}^{+}(\mathbf{v}_z)^k \right] \tag{15.21}$$

$$\mathbf{f}_R^{eq}(\mathbf{z}) = \sum_{i=1}^{n_R} \sum_{k=1}^{m_R} \left[a_{j,k,i}^{R,eq} r_{j,i}^{-}(\mathbf{u})^k + b_{j,k,i}^{R,eq} r_{j,i}^{+}(\mathbf{u})^k \right] \tag{15.22}$$

$$\tag{15.23}$$

and the equivalent coefficients are obtained as

$$a_{j,k,i}^{R,eq} = a_{j,k,i}^{R} + \sum_{p=k+1}^{m_R} a_{j,p,i} \binom{p}{k} \left\langle r_{j,i}^{-}(\psi)^{p-k} \right\rangle \tag{15.24}$$

$$b_{j,k,i}^{R,eq} = b_{j,k,i}^{R} + \sum_{p=k+1}^{m_R} b_{j,p,i} \binom{p}{k} \left\langle r_{j,i}^{+}(\psi)^{p-k} \right\rangle \tag{15.25}$$

The above expressions can also be used for obtaining the equivalent coefficients $a_{j,k,i}^{D,eq}$ and $b_{j,k,i}^{D,eq}$ by replacing ψ with $\mathbf{v}_\psi$ and the upper index R with D.

In the general case the terms $\langle \mathbf{f}_R(\psi) \rangle_\tau$ and $\langle \mathbf{f}_D(\mathbf{v}_\psi) \rangle_\tau$ are different than zero. The averages will correspond to additional static forces in the equations governing the slow motion. The HF motion contributes to the effective stiffness and the damping forces. When the differences $r_{j,i}^{\pm}(\cdot)$ are odd functions with respect to a point in the interval $(\tau, \tau + 2\pi]$, the average of the odd powers $\left\langle r_{j,i}^{\pm}(\cdot)^{2k+1} \right\rangle$, $k = 0, 1, 2, \ldots$ is zero. Furthermore, if the restoring or the damping forces are odd functions with respect to the relative displacements of velocities, then the static contributions to the equations governing the slow motion are zero, as well as the coefficients in front of the even powers of $r_{j,i}^{\pm}(\cdot)$.

In order to evaluate the coefficients in the equations governing the slow motion, an estimate of the HF response is necessary. The equations governing the HF response are obtained by inserting (15.11) into (15.1) and subtracting (15.14):

$$\frac{d^2\psi}{dt^2} + \mathbf{f}_D\left(\frac{d\psi}{dt}\right) - \left\langle \mathbf{f}_D\left(\frac{d\psi}{dt}\right)\right\rangle_\tau$$

$$+\mathbf{f}_D\left(\psi\right) - \left\langle\mathbf{f}_D\left(\psi\right)\right\rangle_\tau - \mathbf{V}\left(t_s\right) = \mathbf{f}_{\mathrm{HF}} \qquad (15.26)$$

where $\mathbf{V}\left(t_s\right)$ represents the last sums in equations (15.15) and (15.16). The two systems of equations (15.26) and (15.14), along with (15.11), are completely equivalent to the original system (15.1), and their solution is not easier. Further simplifications can be obtained by making assumptions about the magnitudes of the response. The following two expressions for the fast motion derivatives can be inserted into (15.26).

$$\frac{d}{dt}\psi = \omega\dot{\psi}(t_s,\tau) + \Omega\psi'(t_s,\tau) \qquad (15.27)$$

$$\frac{d^2}{dt^2}\psi = \omega^2\ddot{\psi}(t_s,\tau) + 2\Omega\omega\dot{\psi}'(t_s,\tau) + \Omega^2\psi''(t_s,\tau) \qquad (15.28)$$

The resulting equation is divided by Ω^2, and after rearranging the terms the following expression is obtained:

$$\psi'' + \frac{1}{\Omega^2}\left[\mathbf{f}_D\left(\frac{d\psi}{dt}\right) - \left\langle\mathbf{f}_D\left(\frac{d\psi}{dt}\right)\right\rangle_\tau\right] + \frac{1}{\Omega^2}\left[\mathbf{\Phi}\left(\psi\right) - \left\langle\mathbf{\Phi}\left(\psi\right)\right\rangle_\tau\right]$$

$$+ \frac{\omega^2}{\Omega^2}\ddot{\psi} + 2\frac{\omega}{\Omega}\dot{\psi}' - \frac{1}{\Omega^2}\mathbf{V} = \frac{1}{\Omega^2}\mathbf{f}_{HF}\left(t_s,\tau\right) \qquad (15.29)$$

The order of the derivatives of the fast motion with respect to the slow and the fast time is assumed to be the same:

$$O\left(\dot{\psi}\right) \sim O\left(\psi'\right) \sim O\left(\dot{\psi}'\right) \sim O\left(\ddot{\psi}\right) \sim O\left(\psi\right) \qquad (15.30)$$

Waves with frequency ω propagating along the chain have a relatively long wave length compared to a fast wave with frequency Ω. Therefore, the differences $r_j\left(\mathbf{z}\right)$ and their powers have relatively small values, and thus the term $\mathbf{V}\left(t_s\right)/\Omega^2$ is small. For a given response this assumption and 15.30 can be checked *a posteriori*. In the case when Ω^2 is much larger than the stiffness to mass ratio, $\Omega \gg \omega$, and there is a lack of resonance effects, the damping and restoring forces, as well as the last few terms in (15.29), can be neglected. Then the so-called inertial approximation [Blekhman, 2000] for the HF motion is obtained:

$$\psi'' = \frac{1}{\Omega^2}\mathbf{f}_{HF}\left(t_s,\tau\right) \qquad (15.31)$$

A discussion on the applicability of the inertial approximation and a comparison with the results obtained by using (15.31) is given in Section 4 and is discussed more generally in [Thomsen 2007].

15.4.1 *Example 1: linear plus cubic non-linearity*

Here the expression for the restoring force for mass j has the following form:

$$f_j^R(\mathbf{u}) = a_{j,1,1}^R r_{j,1}^-(\mathbf{u}) + b_{j,1,1}^R r_{j,1}^+(\mathbf{u})$$
$$+ a_{j,3,1}^R r_{j,1}^-(\mathbf{u})^3 + b_{j,3,1}^R r_{j,1}^+(\mathbf{u})^3 \tag{15.32}$$

The resulting mechanical model corresponds to the Fermi-Pasta-Ulam chain used in computational physics, e.g. [Budinsky and Bountis, 1983]. The equivalent restoring force for the equation governing the slow motion is obtained by using (15.24):

$$\left\langle f_j^R(\mathbf{u}) \right\rangle_\tau = a_{j,1,1}^{R,eq} r_{j,1}^-(\mathbf{z}) + b_{j,1,1}^{R,eq} r_{j,1}^+(\mathbf{z})$$
$$+ a_{j,3,1}^{R,eq} r_{j,1}^-(\mathbf{z})^3 + b_{j,3,1}^{R,eq} r_{j,1}^+(\mathbf{z})^3 + \left\langle f_j^R(\boldsymbol{\psi}) \right\rangle_\tau \tag{15.33}$$

$$a_{j,1,1}^{R,eq} = a_{j,1,1} + 3a_{j,3,1} \left\langle r_{j,1}^-(\boldsymbol{\psi})^2 \right\rangle_\tau \tag{15.34}$$

$$a_{j,3,1}^{R,eq} = a_{j,3,1}^R \tag{15.35}$$

The coefficients $b_{j,k,1}^{eq,R}$ are obtained in a similar way. It is assumed that (15.7) is fulfilled, and hence the coefficients in front of the even powers of $r_j^\pm(\mathbf{z})$ will be zero. As it appears, for the slow motion, the HF response results in an increase or a decrease of the linear stiffness of the springs between the masses, depending on the sign of $a_{j,3,1}$. The change is proportional to the degree of the non-linearity and the square of the HF amplitude. Similar results for a continuous system are obtained in [Thomsen and Blekhman, 2007]. The high order non-linear terms in the restoring force are not affected.

15.4.2 *Example 2: essentially non-linear damping and restoring forces*

The coefficients in (15.33), $a_{j,1,1}^R$ and $b_{j,1,1}^R$, are assumed to be zero. The equivalent coefficients become

$$a_{j,1,1}^{R,eq} = 3a_{j,3,1} \left\langle r_{j,1}^-(\boldsymbol{\psi})^2 \right\rangle_\tau \tag{15.36}$$

$$b_{j,1,1}^{R,eq} = 3b_{j,3,1} \left\langle r_{j,1}^+(\boldsymbol{\psi})^2 \right\rangle_\tau \tag{15.37}$$

The HF response results in a significant change of the system behaviour. Slow waves with small amplitude will propagate along the HF excited chain in a similar way as in a linear chain with stiffness given by (15.36) and

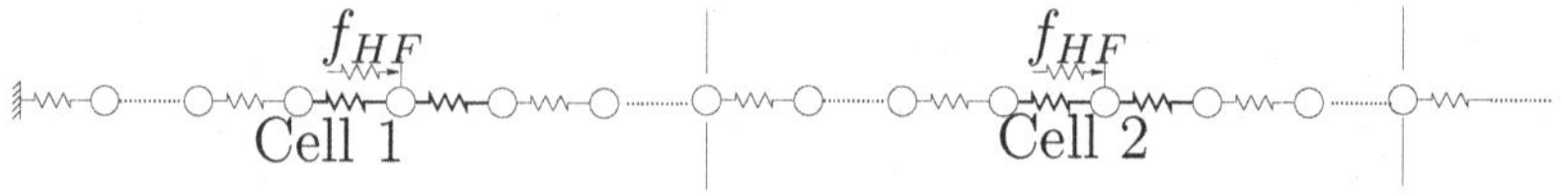

Fig. 15.3   Part of spring mass system with non-linear inclusions.

(15.37). If the damping force for mass j has the same form as (15.33), and the $a_{j,1,1}^D$ and $b_{j,1,1}^D$ are assumed to be zero, HF excitation will add linear viscous damping for slow waves.

15.5   Numerical examples

Numerical simulations are performed with a finite spring-mass chain, using standard numerical integration routines from the SUNDIALS library [Hindmarsh et al., 2005], to test the above analytical predictions. Alteration of the chain stiffness leads to a change of the wave velocity along it. The stiffening effect can be observed by registration of these changes for a system with applied HF excitation. The slow motion has a relatively large wave length, and the contribution of the high-order displacement differences to the restoring force can be neglected. Thus, the wave propagation in the chain with HF excitation can be compared with the wave propagation in an equivalent linear system with parameters estimated by using equations (15.24) and (15.25). For shorter waves with large amplitude it might be necessary to take into account the high order terms in the equivalent restoring force for the slow motion, however, these cases are outside of the scope of this chapter. The numerical tests are performed for stiffening non-linearity. In the case of a softening, a decrease of the wave velocity can be observed. Depending on the HF response amplitude and the type of the softening non-linearity, the HF excitation may also result in destabilisation of the equilibrium position.

The investigated system is a chain with 2000 masses. At both ends the displacements are set to be zero. In every cell of twenty springs and twenty masses, one pair of neighbour springs is elastically nonlinear, while the others are linear (Fig. 15.3). The HF response is generated by harmonic excitation with amplitude raising slowly from zero to the final value A_{HF}.

$$f_{HF} = (1 - \exp(t/C)) \, A_{HF} \sin (\Omega t) \qquad (15.38)$$

Initially, large damping is introduced in the system, as dash-pots between the masses, and the damping coefficients decrease slowly to a pre-scribed small value with the same rate as the HF amplitude. In this way the transient HF response is damped faster, and the contribution of the damping forces to the dynamic equilibrium, when a steady state is reached, is small. The fast decay of the transients helps the system to reach a steady state oscillatory motion, rather than a complex quasi-periodic or chaotic response. Once a steady state due to the HF excitation is reached, a Gaussian modulated pulse is generated by prescribing the displacement at the left end of the chain:

$$u_0(t) = \exp\left(-\alpha\left(t - t_g\right)\right) A_w \cos\left(\omega t - t_g\right) \tag{15.39}$$

where $t_g = 3500 + 3T$, $T = 2\pi/\omega$, A_w is the pulse amplitude, and ω is the slow frequency. The wave velocity for the slow motion can be estimated by finding the displacement maximum $\langle u_j \rangle^{max}$ and $\langle u_i \rangle^{max}$, and by using the following expression:

$$v_w = \frac{j - i}{t_j - t_i} \tag{15.40}$$

where t_j and t_i are the time instances, when mass j and mass i reach the maximum value of the displacements. For the equivalent linear undamped system the group wave velocity can be calculated as a derivative of the dispersion relation $\nu\left(\omega\right)$ with respect to ω, where ν is the phase velocity of a harmonic wave with frequency ω [Brillouin, 1953].

15.5.1   *Inclusions with linear plus cubic non-linear behaviour*

The stiffening effect is illustrated in Fig. 15.4, where snapshots of the spatial displacements of two chains, with and without HF excitation, are presented. A Gaussian modulated pulse with frequency $\omega = 0.025$ is generated by using (15.39) in both chains. The pulse in the chain with HF excitation propagates faster than the one in the chain without. The small spikes in the system with HF excitation are due to the localised HF response in the non-linear inclusions.

A plot of the wave velocity versus the amplitude of the HF excitation, for a chain with non-linear inclusions with restoring force given by (15.32) is shown in Fig. 15.5. An increase of the wave velocity, and hence, an increase of the equivalent stiffness, is observed, which is in line with the analytical prediction. The velocity obtained by numerical simulations for the non-linear system coincides very well with the one obtained from numerical

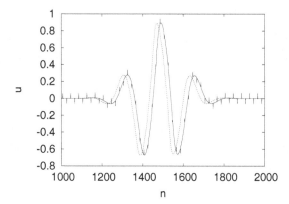

Fig. 15.4   Displacement response for a part of a chain with 2000 masses at time $t = 6400.4$ with applied HF excitation (solid line) and without HF excitation (dashed line).

simulations and analytical calculations for the equivalent linear system. The equivalent parameters are estimated by using (15.34) with HF response obtained from the numerical simulations.

If the restoring and damping forces in (15.29) can be considered to be small compared to the inertial force, the inertial approximation (15.31) for the HF motion is obtained. An estimate of the wave velocity by using the inertial approximation and numerical simulation with the non-linear system is shown in Fig. 15.6. As the HF frequency increases, the inertial approximation gives better results. For lower frequencies the estimation starts to deviate from the numerical simulations when the amplitude of the excitation increases. Increasing the amplitude of the excitation leads to an increase of the amplitude of the HF response, and thus, the restoring force term in (15.29) has a large contribution to the force balance. Therefore, for strongly non-linear systems, the inertial approximation will not lead to an accurate prediction, even in the case when resonance effects are avoided. For weakly non-linear systems it leads to accurate results.

15.5.2   *Non-linear inclusions with non-local interaction*

The theoretical predictions for the stiffening are valid also for systems with non-local interaction between the particles. In order to demonstrate qualitatively their validity, numerical experiments are performed using the spring mass system described in the previous subsection with short non-linear in-

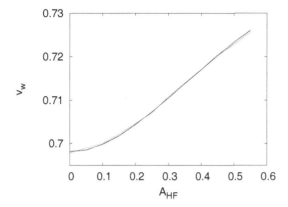

Fig. 15.5   Group wave velocity for a pulse with a base frequency $\omega = 0.025$ in a chain with two non-linear inclusions on every 20 springs, and HF forces (amplitude A_{HF}, frequency $\Omega = \pi$) applied on the masses between the neighbour non-linear springs. The linear springs have stiffness $a_{j,1,1} = 0.5$, and the non-linear springs restoring forces have parameters $a_{j,1,1} = 0.4$ and $a_{j,3,1} = 0.4\gamma$, where $\gamma = 200$. The solid line is obtained by numerical simulations with the non-linear inclusions, and the dashed line - by numerical simulations with the equivalent linear system obtained by replacement of the non-linear springs with equivalent linear ones. The dotted line is obtained for the equivalent system by using analytical prediction for the undamped system.

clusions on every 40 masses. The central particle j interacts with with particles $j + 2$ and $j - 2$ with a purely non-linear force proportional to the relative displacements raised to the power of three. The restoring force for the central mass j in the non-linear inclusion has the following form:

$$f_j^R(\mathbf{u}) = a_{j,1,1}^R r_{j,1}^-(\mathbf{u}) + b_{j,1,1}^R r_{j,1}^+(\mathbf{u})$$
$$+ a_{j,3,1}^R r_{j,1}^-(\mathbf{u})^3 + b_{j,3,1}^R r_{j,1}^+(\mathbf{u})^3$$
$$+ a_{j,3,2}^R r_{j,2}^-(\mathbf{u})^3 + b_{j,3,2}^R r_{j,2}^+(\mathbf{u})^3 \qquad (15.41)$$

The numerical results are shown in Fig. 15.7 for liner stiffness of the springs 0.5, non-linear coefficients $b_{j,3,1} = a_{j,3,1} = 0.5$ and non-local coefficients $b_{j,3,2} = a_{j,3,2} = a_{j,3,1}/\beta$ where $\beta = 1, 2, 4, 8$. The case when $\beta = \infty$, which corresponds to the non-linear forces depending only on the relative displacements between the neighbour masses, is shown with a solid line.

15.5.3   *Linear chains with non-linear damping forces*

HF excitation combined with non-linear damping forces leads to smaller or larger damping for LF waves propagating along the chain. Numerical

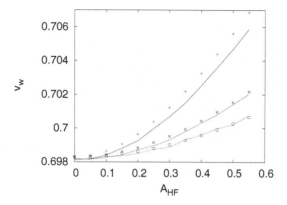

Fig. 15.6   Comparison of the wave speed obtained by using inertial approximation and numerical simulations for the same system as in Fig. 15.5, and HF excitation with frequency $\Omega = 1.5\pi, 1.8\pi, 2\pi$ shown with solid, dashed and dotted lines respectively. Values obtained by the inertial approximation are shown with point markers.

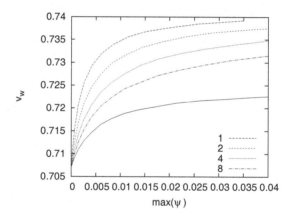

Fig. 15.7   Group wave velocity for a pulse with a base frequency $\omega = 0.025$ in a chain with non-linear inclusions on every 40 springs and non-local interaction forces between the central mass of the inclusions and the rest of the particles.

experiments are performed on a similar setup as in the previous subsections with non-linear damping forces depending on the relative velocities between neighbour masses. In contrast to the case with non-linear restoring forces, the non-linear damping forces are assumed to act on all masses rather than on short inclusions. HF excitation with frequency within the stop band of the undamped chain is applied on every tenth mass. The damping force on mass j is assumed to have the following form

$$f_j^D (\mathbf{u}) = a_{j,1,1}^D r_{j,1}^- (\mathbf{v}) + b_{j,1,1}^D r_{j,1}^+ (\mathbf{v})$$
$$+ a_{j,5,1}^D r_{j,1}^- (\mathbf{v})^5 + b_{j,3,1}^D r_{j,1}^+ (\mathbf{v})^5 \qquad (15.42)$$

where the linear and non-linear coefficients are taken to be equal to 0.0005. A modulated pulse is generated at the first mass using (15.39) with $A_w = 0.1$. The maximal amplitude of the pulse is recorded at mass 400 and the results are shown in Fig. 15.8. As the linear damping is very small, if there is no HF excitation the maximal wave amplitude at mass 400 is almost equal to the initial one. By increasing the amplitude of the HF excitation the wave amplitude decreases, and therefore the effective damping is increasing, in line with the theoretical predictions. The change in the effective damping depends not only on the amplitude of the response but also on the excitation frequency.

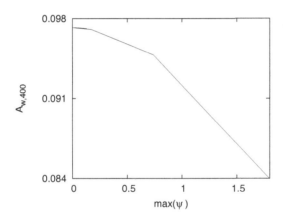

Fig. 15.8   The maximal wave amplitude at mass 400 for different amplitudes of the localised HF response. The excitation frequency is $\Omega = 3\pi$, and the damping force has a linear and a non-linear part, and depends on the relative velocity between the neighbour masses.

15.6   Conclusions

We have demonstrated that HF excitation combined with strong material non-linearities can significantly change the LF wave propagation properties of 1D spring-mass systems. Non-linear restoring forces between the masses result in change of the effective wave speed for long waves, and non-linear damping forces result in change of the effective damping. The effect is proportional to the HF response amplitude for the wave speed, and to the HF frequency and the amplitude for the damping. A stable periodic motion, with large amplitudes of the relative displacements and velocities, may be established by localising the HF response only in parts of the lattice with non-linear behaviour. The localisation is achieved by embedding short non-linear chains into a linear one, and by using HF excitations with frequency within the stop band of the linear chain. Further investigations on the stability of the HF response are necessary, to realise these effects in practical applications. For pulse propagation problems an optimal time-space control strategy could be selected, to minimise the energy required for altering the lattice properties.

Acknowledgments

This work was supported by grant 274-05-0498 from the Danish Research Council for Technology and Production Sciences. The authors wish to thank professor Jakob Soendergaard Jensen for his suggestions and valuable discussions.

References

Blekhman, I. I. (2000) *Vibrational Mechanics - Nonlinear Dynamic Effects, General Approach, Applications*, World Scientific, Singapore

Brillouin, L. (1953) *Wave Propagation in Periodic Structures*, Dover Publications Inc.

Budinsky, N. and Bountis, T. (1983) Stability of nonlinear modes and chaotic properties of 1D Fermi-Pasta-Ulam lattices, *Physica D*, **8**, pp. 445–452

Chatterjee, S. and Singha, T. K. and Karmakar, S. K. (2003) Non-trivial effect of fast vibration on the dynamics of a class of non-linearly damped mechanical systems, *Journal of Sound and Vibration*, **260**, pp. 711–730

Chatterjee, S. and Singha, T. K. and Karmakar, S. K. (2004) Effect of high-frequency excitation on a class of mechanical systems with dynamic friction, *Journal of Sound and Vibration*, **269**, pp. 61–89

Fidlin, A. (2000) On asymptotic properties of systems with strong and very strong high-frequency excitation, *Journal of Sound and Vibration*, **235**, pp. 219–233

Fidlin, A. (2005) *Nonlinear Oscillations in Mechanical Engineering*, Springer-Verlag Berlin

Hindmarsh, A. C., Brown, P. N., Grant, K. E., Lee, S. L., Serban, R., Shumaker, D. E. and Woodward, C. S. (2005) SUNDIALS, suite of nonlinear and differential/algebraic equation solvers, *ACM Transactions on Mathematical Software*, **31**, pp. 363–396

Jensen, J. S. (2000) Buckling of an elastic beam with added high-frequency excitation,*International Journal of Non-Linear Mechanics*, **35**, pp. 217–227

Jensen, J. S., Tcherniak, D. M. and Thomsen, J. J. (2000) Stiffening effects of high-frequency excitation: experiments for an axially loaded beam, *Journal of Applied Mechanics* **67**, pp. 397–402

Kapitza, P. L. (1951) Dynamic stability of a pendulum with an oscillating point of suspension (in Russian), *Zurnal Eksperimental'noj i Teoreticeskoj Fiziki*, **21**, pp. 588–597

Kosevich, Yu. A. (1993) Nonlinear sinusoidal waves and their superposition in anharmonic lattices, *Physical Review Letters*, **71**, pp. 2058–2061

Lazarov, B. S. and Jensen, J. S. (2007) Low-frequency band gaps in chains with attached non-linear oscillators, *Int. Journal of Non-Linear Mechanics* **42**, pp. 1186–1193

Lazarov, B. S. and Thomsen, J. J. (2009) Using high-frequency vibrations and non-linear inclusions to create metamaterials with adjustable effective properties, *Int. Journal of Non-Linear Mechanics* **44**, pp. 90–97

Lazarov, B. S., Snaeland S. O. and Thomsen, J. J. (2008) Using strong nonlinearity and high-frequency vibrations to control effective properties of discrete elastic waveguides, *Proc. of the 6th EUROMECH Conference ENOC 2008* (Saint Petersburg, Russia)

Lurie, K. A. (2000) Low frequency longitudinal vibrations of an elastic bar made of a dynamic material and excited at one end, *Journal of Mathematical Analysis and Applications*, **251**, pp. 364-375

Lurie, K. A. (2007) *An Introduction to the Mathematical Theory of Dynamic Materials*, Springer, New York

Rodriguez-Achahc, M. and Perez, G. (1997) Generalized instability of periodic traveling waves in aharmonic monatomic chains, *Physics Letters A*, **233**, pp. 383–390

Sorokin, S. V., Grishina S. V. and Ershova, O. A. (2001) Analysis and control of vibrations of honeycomb plates by parametric stiffness modulations, *Smart Materials and Structures*, **10**, pp. 1031–1045

Stephenson, A. (1908) On new type of dynamic stability, *Memoirs and Proceedings of the Manchester Literary and Philosophical Society*, **52**, pp. 1–10

Tcherniak, D. M. and Thomsen, J. J. (1998) Slow effects of fast harmonic excitation for elastic structures, *Nonlinear Dynamics*, **17**, pp. 227–246

Thomsen, J. J. (1999) Using fast vibrations to quench friction-induced oscillations, *Journal of Sound and Vibration* **228**, pp. 1079-1102

Thomsen, J. J. (2002) Some general effects of strong high-frequency excitation: Stiffening, biasing and smoothening, *Journal of Sound and Vibration* **253**, pp. 807–831

Thomsen, J. J. (2003) Theories and experiments on the stiffening effect of high-frequency excitation for continuous elastic systems *Journal of Sound and Vibration* **260**, pp. 117–139

Thomsen, J. J. and Blekhman, I. I. (2007) Using nonlinearity and spatiotemporal property modulation to control effective structural properties: Dynamic rods, *In Proc. of ECCOMAS Thematic Conference on Computational Methods, COMPDYN2007*, (Crete, Greece)

Thomsen, J. J. (2007) Effective properties of mechanical systems under high-frequency excitation at multiple frequencies, *Journal of Sound and Vibration*, **311**, pp. 1249–1270

Thomsen, J. J. (2008) Using strong non-linearity and high-frequency vibrations to control effective mechanical stiffness, *Proc. of the 6th EUROMECH Conference ENOC 2008* (Saint Petersburg, Russia)

Yabuno, H. and Tsumoto, K. (2007) Experimental investigation of a buckled beam under high-frequency excitation, *Archive of Applied Mechanics*, **77**, pp. 339–351